INTERDISCIPLINARY TOPICS IN APPLIED MATHEMATICS AND COMPUTATIONAL SCIENCE

INTERDISCIPLINARY TOPICS IN APPLIED MATHEMATICS AND COMPUTATIONAL SCIENCE

Edited by:

Ivan Stanimirović

ARCLER
PRESS

www.arclerpress.com

Interdisciplinary Topics in Applied Mathematics and Computational Science

Ivan Stanimirović

Arcler Press

2010 Winston Park Drive,

2nd Floor

Oakville, ON L6H 5R7

Canada

www.arclerpress.com

Tel: 001-289-291-7705
001-905-616-2116
Fax: 001-289-291-7601
Email: orders@arclereducation.com

ISBN: 978-1-77361-070-2 (Hardcover)

DECLARATION

ABOUT THE EDITOR

Ivan Stanimirovic gained his PhD from University of Niš, Serbia in 2013. His work spans from multi-objective optimization methods to applications of generalized matrix inverses in areas such as image processing and computer graphics and visualisations. He is currently working as an Assistant professor at Faculty of Sciences and Mathematics at University of Niš on computing generalized matrix inverses and its applications.

TABLE OF CONTENTS

List of Contributors xiii

Preface .... *xvii*

Chapter 1 **Monte Carlo Simulation Methodology For Application Sharing And Compatibility: Studies Between Different Radio Services 1**

- Introduction 1
- Monte Carlo Simulation Method: General Considerations 2
- Architectural Requirements 5
- List Of Input Parameters 10
- Generating Device Events 13
- Propagation Model 25
- Propagation Models Combined Insideexterior 32
- Propagation Model For Vhf And Uhf (Itur P.1546) 35
- Test Chi Square Goodness Of Fit 55

Chapter 2 **Alternative Modeling of Applied Mathematics Context In Physics Education: Interdisciplinary Relevance Between Mathematics And Physics 61**

- Introduction 61
- Mathematical Modeling And Symbolic Language 62
- The Theory Of Fields Of Vergnaud Concept: Theoretical Contributions To Mathematical Modeling Applied To Physics Teaching 67
- Qualitative Research And Suggestions For The Development of Mathematical Modeling Applied to Physics Teaching In High School 68
- Final Remarks 73

Chapter 3 **Survey of Interdisciplinary Topics In Computational Science 75**

- Introduction 75

- Technological Training Area ... 82
- Computer Graphics And Image Processing Cg ... 92
- Formulation Of Humanistic Computer Science History ... 95
- Minimum Times For The Courses Of Computer And Information Technology Area ... 108

Chapter 4 Machine Learning Applications in Cancer Prognosis and Prediction..... 109

- Abstract ... 109
- Introduction ... 110
- Ml Techniques ... 111
- Ml And Cancer Prediction/Prognosis ... 119
- Survey Of Ml Applications In Cancer ... 122
- Discussion ... 132
- Conclusions ... 135
- Acknowledgements ... 135
- References ... 136

Chapter 5 Machine Learning And Data Mining Methods In Diabetes Research..... 143

- Abstract ... 143
- Introduction ... 144
- Machine Learning And Knowledge Discovery ... 145
- Diabetes Mellitus ... 149
- Methods ... 150
- Dm Through Machine Learning And Data Mining ... 153
- Discussion ... 165
- Conclusions ... 171
- Acknowledgements ... 172
- References ... 172

Chapter 6 Programmable Genome Editing Tools And Their Regulation For Efficient Genome Engineering ... 189

- Abstract ... 189
- Introduction ... 190
- Genome Editing Reagents ... 192
- Current Regulatable Dna-Cutting Enzymes ... 206
- Alternative Strategies For Developing Regulatable Genome Editing Reagents ... 214

- Conclusion 219
- Acknowledgments 219
- References 220

Chapter 7 A Hybrid Computer-Aided-Diagnosis System For Prediction Of Breast Cancer Recurrence (Hpbcr) Using Optimized Ensemble Learning 243

- Abstract 243
- Introduction 244
- Materials And Methods 248
- Results 256
- Discussion 262
- Conclusions 269
- Acknowledgements 269
- References 269

Chapter 8 The Effects Of Shared Information On Semantic Calculations In The Gene Ontology 281

- Abstract 281
- Introduction 282
- Methods 286
- Results 302
- Discussion 319
- Acknowledgments 322
- Appendix A. Detailed Description Of Shared Information Algorithms 322
- References 325

References 331

Citations 335

Index 337

LIST OF CONTRIBUTORS

Ivan Stanimirović

Konstantina Kourou
Unit of Medical Technology and Intelligent Information Systems, Dept. of Materials Science and Engineering, University of Ioannina, Ioannina, Greece

Themis P. Exarchos
Unit of Medical Technology and Intelligent Information Systems, Dept. of Materials Science and Engineering, University of Ioannina, Ioannina, Greece
IMBB — FORTH, Dept. of Biomedical Research, Ioannina, Greece

Konstantinos P. Exarchos
Unit of Medical Technology and Intelligent Information Systems, Dept. of Materials Science and Engineering, University of Ioannina, Ioannina, Greece

Michalis V. Karamouzis
Molecular Oncology Unit, Department of Biological Chemistry, Medical School, University of Athens, Athens, Greece

Dimitrios I. Fotiadis
Unit of Medical Technology and Intelligent Information Systems, Dept. of Materials Science and Engineering, University of Ioannina, Ioannina, Greece
IMBB — FORTH, Dept. of Biomedical Research, Ioannina, Greece

Ioannis Kavakiotis
Department of Informatics, Aristotle University of Thessaloniki, Thessaloniki 54124, Greece
Institute of Applied Biosciences, CERTH, Thessaloniki, Greece

Olga Tsave
Laboratory of Inorganic Chemistry, Department of Chemical Engineering, Ar-

istotle University of Thessaloniki, Thessaloniki 54124, Greece

Athanasios Salifoglou
Laboratory of Inorganic Chemistry, Department of Chemical Engineering, Aristotle University of Thessaloniki, Thessaloniki 54124, Greece

Nicos Maglaveras
Institute of Applied Biosciences, CERTH, Thessaloniki, Greece
Lab of Computing and Medical Informatics, Medical School, Aristotle University of Thessaloniki, Thessaloniki 54124, Greece

Ioannis Vlahavas
Department of Informatics, Aristotle University of Thessaloniki, Thessaloniki 54124, Greece

Ioanna Chouvarda
Institute of Applied Biosciences, CERTH, Thessaloniki, Greece
Lab of Computing and Medical Informatics, Medical School, Aristotle University of Thessaloniki, Thessaloniki 54124, Greece

Tuhin Kumar Guha
Department of Microbiology, University of Manitoba, Winnipeg, Manitoba R3T2N2, Canada

Alvan Wai
Department of Microbiology, University of Manitoba, Winnipeg, Manitoba R3T2N2, Canada

Georg Hausner
Department of Microbiology, University of Manitoba, Winnipeg, Manitoba R3T2N2, Canada

Mohammad R. Mohebian
Biomedical Engineering Department, Engineering Faculty, University of Isfahan, Hezar Jerib St., 81746-73441, Isfahan, Iran

Hamid R. Marateb
Biomedical Engineering Department, Engineering Faculty, University of Isfahan, Hezar Jerib St., 81746-73441, Isfahan, Iran
Department of Automatic Control, Biomedical Engineering Research Center,

Universitat Politècnica de Catalunya, BarcelonaTech (UPC), C. Pau Gargallo, 5, 08028 Barcelona, Spain

Marjan Mansourian
Department of Biostatistics and Epidemiology, School of Public Health, Isfahan University of Medical Sciences, Hezar Jerib St., 81745 Isfahan, Iran

Miguel Angel Mañanas
Department of Automatic Control, Biomedical Engineering Research Center, Universitat Politècnica de Catalunya, BarcelonaTech (UPC), C. Pau Gargallo, 5, 08028 Barcelona, Spain

Fariborz Mokarian
Cancer Prevention Research Center, Isfahan University of Medical Sciences, Isfahan, Iran
Department of Internal Medicine, School of Medicine, Isfahan University of Medical Sciences, Isfahan, Iran

Paul W. Bible
State Key Laboratory of Ophthalmology, Zhongshan Ophthalmic Center, Sun Yat-sen University, Guangzhou 510060, China

Hong-Wei Sun
Biodata Mining and Discovery Section, Office of Science and Technology, Intramural Research Program, National Institute of Arthritis and Musculoskeletal and Skin Diseases, Bethesda, Maryland

Maria I. Morasso
Laboratory of Skin Biology, Intramural Research Program, National Institute of Arthritis and Musculoskeletal and Skin Diseases, Bethesda, Maryland

Rasiah Loganantharaj
Laboratory of Bioinformatics, Center for Advanced Computer Studies, University of Louisiana at Lafayette, Lafayette, Louisiana

Lai Wei
State Key Laboratory of Ophthalmology, Zhongshan Ophthalmic Center, Sun Yat-sen University, Guangzhou 510060, China

PREFACE

Computational science is a discipline concerned with the design, implementation and use of mathematical models to analyze and solve scientific problems. Typically, the term refers to the use of computers to perform simulations or numerical analysis of a scientific system or process. In practical use, it is typically the application of computer simulation and other forms of computation from numerical analysis and theoretical computer science to solve problems in various scientific disciplines.

This book brings together research and review articles at the interface between applied mathematics, numerical computation, and applications of systems – concerned to the physical, biological, social, and behavioral sciences, and includes papers of a computational nature focusing on new algorithms, their analysis and numerical results. This book lists general information about a Monte Carlo simulation methodology in radio. In addition to providing general information, the text also constitutes a specification of the first generation of software SEAMCAT (spectrum engineering advanced Monte Carlo analysis tool) that uses the Monte Carlo method applied to radio cases.

The problem of unwanted emissions, an important factor affecting the efficiency of radio spectrum use is the subject of an in-depth treatment in various forums both internal and external to the European Conference of Postal and Telecommunications Administrations (CEPT). While the need to re-evaluate the limits of unwanted emissions within Section 3 of the Radio Regulations (RR) is supported, it is generally considered to be preferable to use for this purpose a generic method.

One of the many reasons why generic methods are preferred is their ability, a priori, to deal with new systems and communication technologies as they arise. Another reason is that only a generic method can aspire to become the basis of an analytical tool widely recognized.

The tool for radio Monte Carlo simulation described in this report was developed based on the above considerations, in the process of the European Radiocommunications Committee (ERC).

SEAMCAT is the application of a Monte Carlo simulation model for radio, developed by a group of CEPT administrations, members of the European Telecommunications Standards Institute (ETSI) and international scientific bodies. SEAMCAT is a computer program whose object code is public, being distributed by the European Radiocommunications Office (ERO) of the CEPT.

Computing and information technology area courses have as objectives the training of human resources for the technological development of computing (hardware and software) in order to meet society's needs for the application of computer technology in the interests of society and the formation teachers for middle and vocational education. Between the needs of society can be met with the aid of computers can be mentioned: storage of large volumes of information of all kinds and shapes and their recovery in an acceptable time; computing complex mathematical calculations in extremely short time; secure communication, fast and reliable; automation, control and monitoring of complex systems; fast computing repetitive calculations involving large amounts of information; processing images from different sources; games and tools to support teaching, etc.

Application examples are found in the daily routine of companies (computing involving economic, financial and administrative information generated by business activities, industrial and services); in processing images generated by satellites for weather forecasts; in related to health activities (hospitals, doctors' offices and public health agencies); in air traffic control systems; communication via the Internet; in the banking system, etc. The computation is for man an indispensable and fundamental tool in modern life.

In the context of higher education in the field of Information Technology and its generation processes and automation of knowledge, we must consider the importance of curricula that can effectively prepare critical people, active and increasingly aware of their social roles and its contribution to scientific and technological advancement of the country. The social, humanitarian and ethical content of such training must orient curricula to ensure the expansion of human capabilities in close relation to the technical and scientific learning in the field of Computer and Information Technology. It is therefore a higher education in which individuals are also being trained to deal with human and ethical dimensions of knowledge and social relations.

Applications of computing and computer area courses can be composed of four major areas of training:

- Basic training, which comprises the basic principles of the area of computing, computer science, mathematics necessary to define them formally, physics and electricity needed to enable the understanding and design of viable computers technical and pedagogical training that introduces the basic knowledge the construction of knowledge, necessary for the development of the practice of computer education.
- technological formation applying the basic technological development in computing
- additional training allowing an interaction of the graduates of the courses with other professions

Editor

1 MONTE CARLO SIMULATION METHODOLOGY FOR APPLICATION SHARING AND COMPATIBILITY: STUDIES BETWEEN DIFFERENT RADIO SERVICES

Ivan Stanimirović

INTRODUCTION

In order to proceed to a new assessment of the limits of unwanted emissions, we should develop an analytical tool to assess the level of interference would experience representative receivers. It is agreed in the ITU-R interference level be expressed in terms of the probability that the capacity of the receiver considered degradation by the presence of an interfering signal. For this statistical probability of interference pattern of interference situations you will be needed. This Report describes the methodology and a proposed architecture of the tool is presented.

Statistical methodology described and was used to develop the tool known as Monte Carlo technique. The term Monte Carlo was adopted by von Neumann and Ulan during World War II as codename of its secret work resolution statistical design issues of an atomic bomb. Since then we have used the Monte Carlo method to simulate random processes. It is based on the principle of sampling random variables of their probability density functions defined. This method can be considered the most powerful and most widely used for the analysis of complex statistical technique problems. It is considered that there is no alternative to the Monte Carlo approach to develop a methodology for the analysis of the interference caused by unwanted emissions.

The approach is:

- Generic: a single model can handle a variety of situations of interference;
- Flexible: the approach is highly flexible and can adaptarse easy to deal with situations composite interference.

MONTE CARLO SIMULATION METHOD: GENERAL CONSIDERATIONS

This methodology is appropriate to address the following issues radio spectrum engineering:

- is yourgod sharing and compatibility between different radio systems which operate respectively in the same frequency band or in adjacent bands;
- assessment templates transmitter and receiver;
- evaluation of límitis parameters such as levels of unwanted signals (spurious and OOB), blocking levels and intermodulation levels.

The Monte Carlo method can be applied to almost all radio frequency interference situations. Flexibility is achieved by the way that system parameters are defined. The input format of each of the variables (antenna pattern, radiated power, propagation path, etc.) parameters is the statistical distribution function. This enables models to create very complex situations using relatively simple elementary functions. They can be treated several different systems, such as:

- broadcasting (terrestrial and satellite);
- mobile (terrestrial and satellite);
- point by point;
- point to multipoint, etc.

The best way to explain the principle is the following example, which only takes into account unwanted emissions as interference mechanism.

Typically, the Monte Carlo method is also other effects present in the radio, such as out of band emissions, the receptor blocking and intermodulation environment.

Here are some examples of applications of this methodology:

- compatibility study between digital private mobile media (PMR) (TETRA) and the Global System for Mobile Communications (GSM) 915 MHz;
- sharing study between FS and FSS;
- sharing study between short range devices (Bluetooth) and radio local area networks (RLAN) in the band for industrial, scientific and medical (ISM) 2.4 GHz;
- compatibility study for International Mobile Telecommunications2000 (IMT2000) and PCS1900 around 1.9 GHz;
- compatibility study for ultra-wide band systems and other radio systems operating in these frequency bands.

Illustrative example (only unwanted emissions, main interference source)

It has been assumed that, to consider interference, should not satisfy the minimum carrier / interference, C / I, at the receiver input ratio. To calculate the value of C / I experienced by the receiver, it is necessary to obtain statistics of the levels of desired signal and undesired signal. It is assumed that unwanted emissions in this simulation come from active transmitters. Furthermore, it has been considered that only contribute to interference Spurious signals falling within the reception bandwidth. In the case of interference from mobile to fixed, an example shown in Fig. 1.

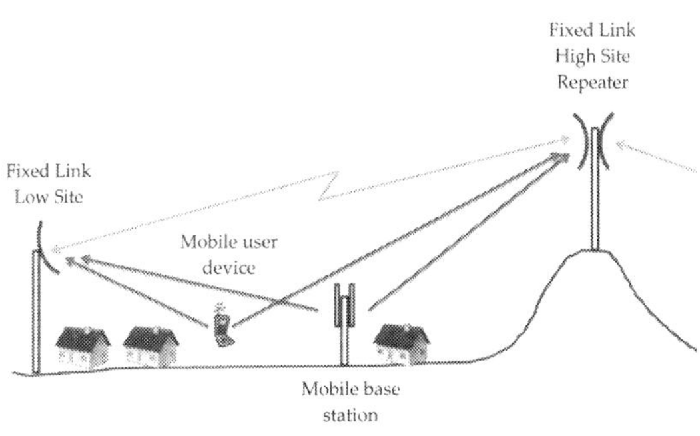

Figure numerous potential mobile transmitters appear. Only some of the transmitters are actively transmitting and only a few emit unwanted bandwidth power from the affected receiver. It is assumed that the interference occurs as a result of unwanted emissions transmitter greater influence and the minimum path loss (median value of the propagation loss + variation additional attenuation + change in transmission power) to the receiver.

An example process Monte Carlo simulation shown in Fig. 2 as to the calculation of the probability of interference from unwanted emissions. For each trial a random draw the desired level of signal distribution is appropriate. For a desired signal level determined maximum tolerable level unwanted receiver input from the value of C / I receptor is obtained.

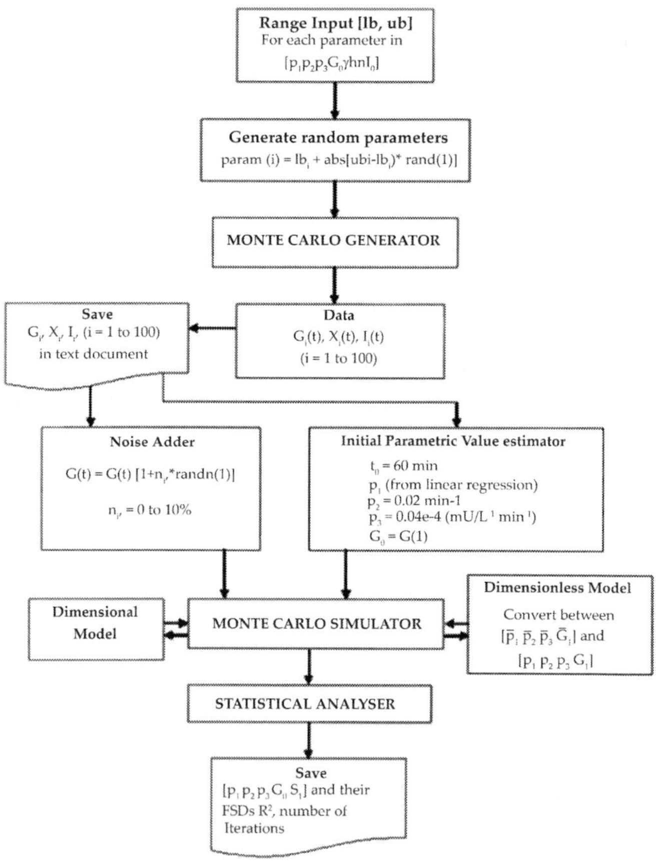

It is calculated, for the different interferers surrounding the affected receiver, the isolation position, the propagation loss (including any variations and additional losses) and antenna discrimination. The lowest isolation determines the maximum level that can radiate unwanted any of the transmitters during this test.

From many tests can then obtain a histogram of unwanted levels and, for a given probability of interference can be determined level corresponding unwanted.

Varying the values of various parameters input to the model and given an appropriate density of interferers, it is possible to analyze a wide range of interference situations

ARCHITECTURAL REQUIREMENTS

One of the main requirements is to choose a structure of architecture for simulation tool flexible enough to allow analysis of situations composite interference, in which involved a combination of radio equipment that share the same habitat and / or more interference sources (for example, out of band emissions, spurious emissions, intermodulation, etc.) which may be treated simultaneously.

Other requirements could be that the proposed architecture consists of modular elements and is versatile enough to allow treating composite interference situations.

In Fig. 3 the Monte Carlo proposed architecture and complies with these restrictions is shown. In essence, the proposed architecture is sequential and consists of four processing devices:

- d generating deviceand events;
- calculating means for evaluating distribution;
- interference calculation device;
- evaluating device limits.

Annex 1 provides a list of interference parameters and their importance with respect to one or more of the processing devices.

Generating device events

The generating device events (EGE, event generation engine) takes the relevant parameters of the interference situation presented and generates information about the intensity of the received signal (RSS, received signal strength) desired as well as the intensity of each of the interfering signals that are part of the composite interference situation. The process is repeated N times, where N is the number of tests which must be large enough to generate statistically significant results. Generated samples of the desired signal and interfering signals from all matrices are stored in separate data length N.

Trials common to the desired and interfering radio path parameters are done concurrently to detect a possible correlation between the wanted and interfering signals. Such an embodiment does not encompass only those rare cases of interference in a certain interference mechanism is excited by another interference (for example, the signal of a powerful transmitter is mixed with the spurious emission of the second transmitter and produces an interference type intermodulation).

The list of possible sources of interference that may appear in a radio environment includes:

Transmitter interference phenomena:

- unwanted emissions (spurious and OOB);
- broadband noise;
- intermodulation;
- adjacent channel;
- cocanal.

Receiver interference phenomena:

- spurious emission.

Background noise:

- noise antenna;
- MMN.

Other parameters receptor susceptibility to interference:

- blocking;
- intermodulation rejection;
- rejects adjacent channel and co-channel;
- rejection response not essential.

All of the above sources can be classified into three categories of interference generic mechanisms: unwanted emission, intermodulation susceptibility receptor. Each of these three categories requires a different model for the physical processes that are characteristic of said interference mechanism. The artificial noise and the noise temperature of the antenna can be considered as an increase in thermal noise level, which decreases the sensitivity of the receiver, and can be introduced into the simulation when the interference criterion is I / N (interference ratio / noise) or C / (I + N) (desired signal / interference + noise).

Evaluation device distribution

The evaluation device distribution (DEE, distribution evaluation engine) data collection matrices generated by the EGE and processes the data to:

a) assess whether or not the number of samples is sufficient to generate statistically stable results;

b) calculatedr the correlation between the data of the desired signal and the interfering signal, as well as between different types of interfering signals (for example, emissions and blocking unwanted emissions);

c) calculating a continuous distribution function conocway, for example Gaussian, as best fit to the data distributions generated desired and interfering signals.

Points a) and c) can be achieved using well known for calculating the goodness of fit for general distributions, such as Kolmogorov-Smirnov test algorithms. It thinks investigate further, in the second phase of development of this methodology, the applicability of adjustment to this specific task.

If the DEE detects an unacceptable variation of the parameters of the discrete distribution obtained in two successive estimates using sample sizes , is ordered to generate EGE •N additional samples. This test is repeated until a tolerable variation of the parameters over a predefined number of successive tests is measured.

The DEE has three different output types possible:

- data arrays wanted and interfering signals. This is the output if a high degree of correlation between the desired signal and any interfering signals detected;

- discrete distributions appear signals deseada and interfering; this occurs when the correlation between the signals is weak or no correlation between the signals but it has not been possible to approximate a continuous distribution with satisfactory precision;
- They transferred to dispositivo calculating interference (ICE, interference calculation engine) continuous distribution functions of the desired and interfering signals, if the signals were uncorrelated and is achieved by approximating the discrete distributions continuous distribution functions.

Annex 3 the proposed flow chart and detailed specification of the algorithm appears.

Calculation device interference (ICE)

ICE is the heart of the proposed architecture. ICE uses the information gathered by the EGE and processed by the DEE to calculate the probability of interference. As shown, depending on the type of information transferred DEE to ICE, we can identify three possible ways of calculating the probability of interference.

mode 1: Data matrices ^{dRSS} (Desired signal) and inRSS (interfering signal coming from n different systems) are transferred by the DEE to ICE and the vector representing the composite signal I interferer is calculated as the sum of vectors inRSS data.

mode 2: The distribution function of the interfering composite signal is calculated by taking random samples and adding them inRSS distributions linearly

mode 3: InRSS is calculated by numerical integration or analytical distribution function given to each of the sources of interference.

mode 4: Considering that all signals are mutually independent and the probability of interference is identified overall probability of appearing disturbances caused by at least one type of interference.

Within the processing device can support different criteria for calculating the probability of interference. Is possible to determine the cumulative probability functions for random variables

C / I,C / (N + I),I / N or N / (N + I).

Flow information, along with associated processes, are shown in flowchart form in Annex 4.

All distributions are calculated interfering signals with respect to reference levels or functions of unwanted attenuation (template emission), lock (template reception) or intermodulation. Interfering signal distributions for levels or other reference functions can be obtained by linear translation of first (unwanted or blocking) order or third order (intermodulation) distributions reference (see Section 4).

Boundary evaluation device (LEE, *Limits evaluation engine*)

Lee plays a very important role in two aspects of development of the tool:

- selection of optimum values for limits;
- verification tool.

ICE output appears as a multidimensional surface where the probability of interference based on radio parameters shown. Two of the main features of the probability surface are:

- equal probability of interference is obtained with different sets of limit values of the considered radio parameters;
- parameter probabilisticbility interference is not used in the design of radiocommunication systems and, therefore, does not readily lend validation by measurements of the quality of system performance. Instead, degradation of system coverage or traffic capacity are most appropriate parameters to understand the impact of a given probability of interference on the performance of the radio system.

Radio variables become the probability space into a space of system performance, thus allowing to evaluate the quality degradation operating due to the presence of interference system. When compatibility between systems is analyzed (eg unwanted emissions) may used radio coverage and / or traffic capacity to evaluate the influence of the boundaries of the radio parameters. In the case of a compatibility study within the system (eg outband emission), use efficiency in spectrum utilization for the appropriate radio parameters.

The limit values are determined by means of an optimization algorithm. Optimization to work properly, you must set a criterion that is generally called cost and process optimization function aims to minimize this cost function. It is a feature of all radio parameters and their meaning for the cost, it can be modified using weights.

The weights can be integrated into the optimization process any of the following:

- system availability;
- traffic capacity;
- spectrum utilization;
- limitations of technology;
- economic constraints.

The set of values of radio parameters for which the cost function is minimized represents the optimal solution of the limit values.

The role played by Lee in the tool is very important. However, as still they are analyzed its various elements, not be possible to include the LEE in this first phase of the embodiment

LIST OF INPUT PARAMETERS

The following rules apply:

- distribution functions are capitalized letter, for example,*P*;
- lowercase represent variables (result of a calculation or trial), for example,*p*;
- the index refers to a "protagonist": tranSmisor desired, affected receiver, receiver and interfering transmitter desired.

Parameters of the desired transmitter (wt):

$P_{wt}^{supplied}$: power level distribution for various transmitters (dBm)

$p_{wt}^{supplied}$: power level sample extracted from the previous distribution (dBm)

$g_{wt}^{máx}$: maximum antenna gain (dBi)

patternwt: antenna directivity within the operating bandwidth (dB) (given as a function or lookup table)

HWT: distribution of the antenna height (1 / m)

$R_{wt}^{máx}$: Radio coverage of the desired transmitter (km) (not required for point to point).

Parameters affected receiver (vr):

C/ I: protection ratio (dB)

$g_{vr}^{máx}$: maximum antenna gain (dBi)

patternvr: antenna directivity within the operating bandwidth (dB) (given as a function or lookup table)

Hvr: distribution of antenna height (1 / m)

block: receiver frequency response (dB)

avr: característica susceptibility receptor; is expressed as a ratio between the levels of unwanted interfering signal producing unacceptable performance receiver and n is a function of frequency separation between the two signals

intermod: receptor response to intermodulation (dB)

The intermodulation response to receptor measures the ability to receive a desired modulated signal without exceeding a given degradation due to the presence of two unwanted signals aa specific frequency relationship with respect to the frequency of the desired signal.

FVR: Frequency (MHz)

sensvr: affected receiver sensitivity (dBm)

BVR: bandwidth affected receiver (kHz).

Interferer parameters (it):

$P_{it}^{supplied}$: power level distribution for various transmitters (dBm)

$p_{it}^{t\_hold}$: power control threshold (dBm)

$p_{it}^{dyc\_rg}$: dynamic range power control (dB)

$p_{it}^{st\_rg}$: Pass interval control power (dB)

$g_{it}^{máx}$: maximum antenna gain (dBi)

$R_{it}^{máx}$: Radio coverage interferer (km)

R_{simu}: radius of the area where interferers (km) are distributed

d_0: minimum protection distance (km) between the victim receiver and the interfering transmitter

patternit: antenna directivity (dB) (given as a function or lookup table)

emission_relit: template relative emission (dBc / (bandwidth reference)), used only for the interferer and consisting of the desired signal level and all unwanted emissions, including part of the minimum emission threshold depending on the power control

emission_floorit: minimum threshold absolute emission (dBm / (reference bandwidth)), used only for the interferer (unwanted emissions to be issued with the minimum possible transmitter power).

fit: Frequency (MHz)

densit: density (1 km^2)

p_{it}^{tx}: transmission probability (%); It is a statistical description of the activities averaged transmitters between a large number of users and over a prolonged period of time

Tempit: temporal variation of the normalized activity, depending on the time of day (1 / h) function.

Parameters of the intended recipient (wr) corresponding to the interfering transmitter:

$g_{wr}^{máx}$: maximum antenna gain (dBi)

patternwr: antenna directivity (dB) (given as a function or lookup table)

HWR: distribution of antenna height (1 / m)

senswr: Dynamic des receiver sensitivityOCTA, taking into account a margin for fast fading and interference within the system (dBm).

Environmental parameters and propagation:

fpropag: propagation law (median value of losses

fmedian: propagation law (median value only losses)

env: type (indoor / outdoor, urban / suburban / open field) environment.

GENERATING DEVICE EVENTS

INTRODUCTION

This section describes how to build signals used in interference situations described: the desired signal and interference signals due to unwanted emission to blocking and intermodulation. The calculated signals are stored in a matrix which constitutes the entrance to DEE, as shown in Fig. 4.

Organigrama general del EGE

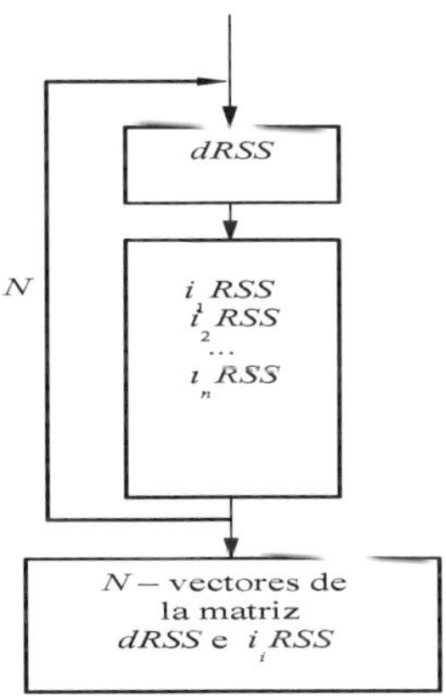

Rap 2028-04

Departures

dRSS: intensity desired received signal (dBm)

iRSSspur: received signal strength including unwanted interference emissions (dBm)

iRSSblocking: received signal strength due to blocking interferer (dBm)

iRSSintermod: Receive signal strengthto interference due to intermodulation (dBm).

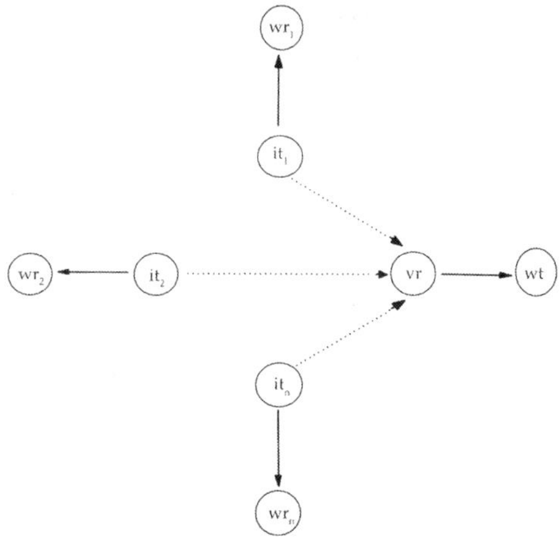

Calculation

In this point:

- *T* It represents a distribution of a given test (the algorithm described in Section 4).
- distributions *OR*(0.1), G (·) and R (•) are defined in Section 3.
- The organizational chart for calculating *dRSS* in Section 5 and flowcharts for calculating SSRIs in Appendices.

NOTE 1 - The distance d between transmitters are expressed in km.

a) Calculation of *dRSS*

There are three different choices to determine *dRSS*: Depending on a variable distance, for a fixed distance or by using a given signal distribution (see Section 5).

If variable distance:

$$dRSS = f\left(p_{wt}^{supplied}, g_{wt\rightarrow vr}, pl_{wt\rightarrow vr}, g_{vr\leftrightarrow wt}\right) = p_{wt}^{supplied} + g_{wt\rightarrow vr}\left(f_{vr}\right) - pl_{wt\rightarrow vr}\left(f_{vr}\right) + g_{vr\leftrightarrow wt}\left(f_{vr}\right)$$

If the received signal can not exceed a given value (ie, if power control depends on the affected system adopted), then:

dRSS• min (dRSS, DRSSmáx) using dRSS as calculated before

where:

FVR: frequency received at the receiver.

$f_{vr} = T(f_{vr})$

This frequency can be set at a constant value or determined to a certain distribution, the "distribution of discrete frequencies". In general, the affected frequency should not be set to a value but must be calculated and randomly chosen as the frequency of the interferer using a discrete distribution).

$p_{wt}^{supplied}$: distribution desired level of maximum power supplied to the transmitter antenna

$p_{wt}^{supplied} = T\left(p_{wt}^{supplied}\right)$

$pl_{wt\leftrightarrow vr}$: Path loss between the transmitter and desiredthe affected receiver (taking into account propagation loss, slow fading and clutter losses). Depending on whether the interference criterion is applied to the instantaneous dRSS (Rayleigh fading excluded) or the average value of dRSS:

$pl_{wt \to vr} = f_{propag}\left(f_{vr}, h_{vr}, h_{wt}, d_{wt \to vr}, env\right)$

or

$pl_{wt \to vr} = f_{median}\left(f_{vr}, h_{vr}, h_{wt}, d_{wt \to vr}, env\right)$

where:

hvr: height victim receiver antenna

$h_{vr} = T(H_{vr})$

HWT: desired height of the transmitter antenna

$h_{wt} = T(H_{wt})$

$d_{wt\leftrightarrow vr}$: distance between the affected receiver and transmitter desired

$d_{wt \to vr} = T\left(R_{max}^{wt}\right)$

for example: $d_{wt \to vr} = R_{max}^{wt}\sqrt{T\left(U(0,1)\right)}$

three different options are taken into account R_{max}^{wt} :

Option 1: given distance R_{max}^{wt} .

Option 2: Noise limited network.

R_{max}^{wt} It is determined by the following equation:

$$f_{median}\left(f_{vr}, h_{vr}, h_{wt}, d_{wt\rightarrow vr}, env\right) + f_{slowfading}\left(X\%\right)$$

where:

$F_{mediate}$: propagation loss not including slow fading

$F_{slowfading}$(X%): fade margin to be used for a loss of coverage of 1-X%.

In the case of lognormal fading distribution and a loss of 95% coverage in the edge area cobertura, the value of fslowfading for long distances is well known: 1,64 times the standard deviation of the propagation loss. Section 11 details on determining the size of radio cell in a noise limited network appear.

option 3: Limited network traffic

$$R_{max}^{wt} = \sqrt{\frac{n_{channels} n_{userspecchannel}}{\pi dens_{max} cluster_{frequency}}}$$

GWT Transmitter antenna gain desired direction affected receiver

$$g_{wt\rightarrow vr} = f\left(g_{wt}^{max}, pattern_{wt}\right) = g_{wt}^{max} \times pattern_{wt}\left(\theta_{wt\rightarrow vr}, \varphi_{wt\rightarrow vr}, f_{vr}\right)$$

where: angles azimuth and elevation between the top of the antenna of the desired transmitter and the top of the victim receiver antenna

for example:

$$\theta_{wt\rightarrow vr} = T(U\left(0, 2\pi\right) = 2\pi T\left(U\left(0,1\right)\right)$$

$$\varphi_{wt\rightarrow vr} = T(U\left(-\pi/2, \pi/2\right) = \pi T\left(U\left(0,1\right)\right)_{-} \pi/2$$

gvr gain of victim receiver antenna towards the transmitter desired

$$g_{vr\rightarrow wt} = f\left(g_{vr}^{max}, pattern_{vr}\right) = g_{vr}^{max} \times pattern_{vr}\left(\theta_{wt\rightarrow vr} + \pi, -\phi_{wt\rightarrow vr}, f_{vr}\right)$$

If fixed distances:

$P_{wt}^{nominal}$: rated power distribution

$f_{fading,\ fixed\ link}$: fading distribution

$$dRSS = f\left(P_{wt}^{nominal}, f_{fading,fixedlink}\right) = T\left(P_{wt}^{nominal}\right) - T\left(f_{fading,fixedlink}\right)$$

If given dRSS: Distribution determined by the user.

b) calculation iRSSblock

$$IRSS_{block} = \sum_{j=1}^{ninterferers} f(p_{it}^{supplied}, g_{it}^{PC}, g_{it\to vr}, pl_{it<\to vr,a_{vr}}, g_{vr\to it})_j = 10\log \sum_{j=1}^{ninterferers} 10^{iblockj/10}$$

where the jth signal is interfering given by:

$$iblockj = (p_{it}^{supplied} + g_{it}^{PC} + g_{it\to vr} - pl_{it<\to vr,a_{vr}} + g_{vr\to it})_j$$

and for each interferer:

F_{Item}: transmission frequency interferer

$$f_{it} = T\left(f_{it}\right)$$

For the "distribution of discrete frequencies' see Section 3.

Note that it is clear that the test frequency *dRSS*It RVF occurs once and only once in each round of simulation; ie fvr tested once affected as desired positions, the desired transmitter power and other distributions on the affected link. These distributions tested dRSS values apply to <N tests IrSS (N being the number of interferers).

If I could limit the randomness of some parameters, the model would not be used only for simulation but also to make more accurate calculations. This feature PermiTyrian more easily verify the validity of simulation results.

$p_{it}^{supplied}$: Maximum power delivered to the interfering transmitter antenna (before power control)

g_{ic}^{PC}: Warranty of power control interfering transmitter

$$g_{it}^{PC} = f_{pc}\left(p_{it}^{supplied}, g_{it\to vr}, pl_{it\leftrightarrow vr}, g_{vr\to it}, pc_{it}^{t\_hold}, pc_{it}^{dyc\_rg}, pc_{it}^{st\_rg}\right)$$

where:

F_{pc}: power control function (given in Section 2)

$pl_{Item«wr}$: path loss between the interfering transmitter and the intended receiver (taking into account propagation loss, slow fading and clutter losses). Depending on the embodiment of the power control can tratarsand the average path loss or instantaneous path loss (excluding Rayleigh fading)

or

where:

h_{wr}: desired height of the transmitter antenna

for example:

hit: height interfering transmitter antenna

$h_{it} = T(H_{it})$

for example:

$d_{it\leftrightarrow wr}$: distance between the interfering transmitter and the desired transmitter

)

for example:

three different options are chosen R^{it}_{max}:

Option 1: given distance R^{it}_{max}

Option 2: Noise limited network

option 3: Limited network traffic

In section a) it is shown the details on the determination of cell size.

$git_{®wr}$: antenna gain of the interfering transmitter in the direction of the nearest base station

$$g_{wr\rightarrow it} = f\left(g^{max}_{wr}, \text{pattern}_{wr}\right) = g^{max}_{wr} \times \text{pattern}_{wr}(\theta_{it\rightarrow wr} + \pi, \phi_{it\rightarrow wr}, f_{it})$$

where:

$(\theta_{it\rightarrow wr}, \phi_{it\rightarrow wr})$: angles azimuth and elevation between the top of the interfering transmitter antenna and the top of the receiver antenna desired for example: $\theta_{wt\to vr} = T(U(0,2\pi) = 2\pi T\left(U\left(0,1\right)\right)$

$\varphi_{wt\to vr} = T(U\left(-\pi/2,\pi/2\right) = \pi T\left(U\left(0,1\right)\right)_{-}\pi/2$

To calculate the gain symmetric antenna diagrams see Section 12.

$gwr_{®i\tau}$: antenna gain of the base station in diRecciôn the interfering transmitter

$$g_{wr} \to it = f\left(g_{wr}^{max}, pattern_{wr}\right) = g_{wr}^{max} xpattern_{wr}\left(\theta_{wt\to vr} + \pi - \varphi_{wt\to vr}, fit\right)$$

$pl_{it\leftrightarrow vr}$: path loss between the interfering transmitter i and the affected receiver (taking into account the propagation loss, slow fading and clutter losses)

$$pl_{it\to vr} = fpropag\left(f_{it}, h_{vr}, h_{it}, d_{it\to vr}, env\right)$$

or

$$pl_{wt\to vr} = fmedian\left(f_{it}, h_{vr}, h_{it}, d_{wt\to vr}, env\right)$$

The choice between *fmedian*and fpropag depend on the interference criteria and is closely related to the choice made for estimating dRSS; ic if the calculation made by ICE out:

$$\frac{dRSS_{mean}}{iRSS_{mean}}, \frac{dRSS_{propag}}{iRSS_{propag}}, \frac{dRSS_{mean}}{iRSS_{propag}}$$

where:

hvr: height victim receiver antenna (defined in calculation *dRSS*)

hit: height interfering transmitter antenna (previously defined)

$d_{it\leftrightarrow vr}$: distance between the victim receiver and the interfering transmitter.

There are three different ways to choose $d_{it\leftrightarrow vr}$:

1. The most common case is one in which there is no spatial correlation between the elements of the affected system and elements of the interfering system.

So, $d_{it\leftrightarrow vr}$ It is the result of a test:

$$d_{it\to vr} = R_{simu}\sqrt{T\left(U\left(0,1\right)\right)}$$

where:

Rsimu: radius of the area where interferers are spread

$$R_{simu} = \sqrt{\frac{n^{active}}{\pi densit^{active}}}$$

where:

n active: number of active interferers considered in la simulation

n active: It should be large enough to interferer n 1 to provide extra power negligible interference

$$dens_{it}^{active} = dens_{it} \times p_{it}^{ts} \times temp(time)$$

If introduced minimal protection of $\Delta/T_{«vr} \geq d_0$ between the victim receiver and the interfering transmitter, Rsimu is calculated as follows:

$$R_{simu} = \sqrt{\frac{n^{active}}{\pi densit^{active}} + d_o^2}$$

Note that if the distance protection d0 0 a uniform distribution of the interfering transmitter should be chosen.

2. In the present case examines the situation in which the affected system and interfering system are geographically correlated (ie, base stations placed on the same site).

It is assumed that this correlation only between an element (transmitter desired or affected) the affected system and an element (desired transmitter or interference) of the interfering system.

a test is made (if the distance is not fixed) of the distances and angles between the two correlated elements (for example, $d_{wr\leftrightarrow vr}, \theta_{wr\leftrightarrow vr}$). Known values $d_{it\leftrightarrow wr}, d_{vr\leftrightarrow wt}, \theta_{it\leftrightarrow wr}, \theta_{vr\leftrightarrow wt}$ can be calculated missing (e.g. coordinates, $d_{it\leftrightarrow vr}, \theta_{it\leftrightarrow vr}$).

Situación de interferencia con correlación geográfica entre el sistema afectado y los sistemas interferentes

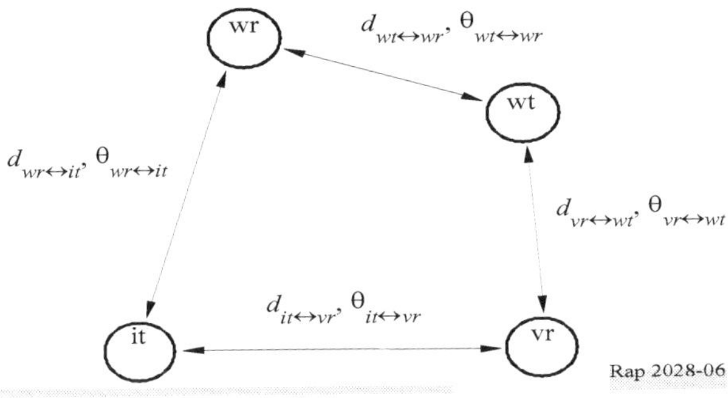

3. Source interfering at nearest

The influence of the interference source more next it can be estimated by setting a distance of DIT follow a Rayleigh distribution R defined in Section 3. It is related to the density of transmitters. This is an alternative method to calculate the relative location of the interfering transmitter relative to the receiver affected uncorrelated manner which avoids perform multiple tests on the number of interferers.

In this case, the distribution for the distance between *Item* Y *vr* in the simulation it is always a Rayleigh distribution:

$$d_{it\to vr} = R_{simu} \times R(\sigma)$$

It is related to the density of active transmitters:

$$\sigma = \frac{1}{\sqrt{2\pi dens_{it}^{active}}}$$

Note that the simulation radius is useless but associated parameters (density, activity and probabilisticness) are still needed to calculate the density of active transmitters.

$$dens_{it}^{active} = dens_{it} \times pit \times activity$$

$g_{it\to vr}(f_{it})$: antenna gain of the interfering transmitter in the direction of the affected receiver

$$g_{it\to vr} = f\left(g_{it}^{max}, pattern_{it}\right) = g_{it}^{max} \times pattern_{it}(\theta_{it\to vr}, \varphi_{wt\to vr})$$

where:

$(\theta_{it\to vr}, \phi_{it\to vr})$: azimuth angles and elevation between the top of the antenna closest interfering transmitter and the top of the antenna of the affected receiver

for example: $\theta_{it\to vr} = T(U(0,2\pi) = 2\pi T(U(0,1))$

$\varphi_{it\to vr} = T(U(-\pi/2,\pi/2) = \pi T(U(0,1))_{-}\pi/2$

avr(Fit, FVR): attenuationng the receiver.

three possible ways to calculate this attenuation are taken into account:

1. *avr* It is given by the user.
2. The blockade is expressed in terms of blocking attenuation ratio or protection. For a desired 3 dB signal is above the sensitivity, you can obtain the attenuation avr from the following equation (see Section 7):

$$a_{vr} = f\left(\frac{C}{N+1}, blockatt\right) = 3 + \frac{C}{N+1} + \text{blockatt}\,(f_{it}, f_{vr})$$

The blockade is expressed in terms of the absolute level of blocking:

$$a_{vr} = f\left(\frac{C}{N+1}, blockatt\right) = 3 + \frac{C}{N+1} + \text{blockatt}(f_{it}, f_{vr}) - sens_{vr}$$

It PUEden provide two cases:

Case 1: *block* It is a template that is a function of $\Delta f = (f_{it} - f_{vr})$. It is introduced to calculate interference between adjacent band systems.

case 2: *block* It is a fixed value (for example, 80 dBm). It is used to calculate generic limits.

$g_{vr\to it}(f_{it})$: gain victim receiver antenna in the direction of the interferg transmitter

$$g_{vr\to it} = f\left(g_{vr}^{max}, \text{pattern}_{vr}\right) = g_{vr}^{max} \times \text{pattern}_{vr}(\theta_{it\to vr}, \phi_{it\to vr}, f_{it})$$

c) calculation ***iRSSspur***

$$IRSS_{spur} = \sum_{j=1}^{ninterferers} f(emmision_{it}, g_{it}^{PC}, g_{it\to vr}, pl_{it<\to vr, a_{vr}}, g_{vr\to it})_j = 10\log \sum_{j=1}^{ninterferers} 10^{ispurj/10}$$

where the jth interfering signal is defined as:

$$ispurj/10 = (emmision_{it}\left(f_{it}, f_{vr}\right) + g_{it\to vr}\left(fvr\right) - pl_{it\to vr}\left(f_{vr}\right) + g_{vr\to it}\left(f_{vr}\right)$$

Most parameters are already defined in paragraphs a) or b).

emissionit(Fit, fvr): emission template for the interfering transmitter which generally depends on the template relative emission, interference power, gain control power and bandwidth of the emission increased by the minimum absolute threshold of emission. Section 10 further details and a study of the influence of different band widths of the desired and interfering radio systems appear.

$$\text{emission}_{it}(f_{it}, f_{vr}) = \max\left\{ p_{it}^{supplied} + \text{emission\_rel}_{it}(f_{it}, f_{vr}) + g_{it}^{PC},\ \text{emission\_floor}_{it}(f_{it}, f_{vr})\right\}$$

*emission*_relit: It is a template that is relative emission function • f • (fit, fvr). It is introduced to calculate the interference between systems located in the same band or in adjacent bands. The actual emission is always above or equal to the absolute minimum threshold emission emission_ floorit (fit, fvr).

g_{it}^{pc}: power control gain interfering transmitter (definway in b))

$pl_{it\leftrightarrow vr}$: path loss between the interfering transmitter and receiver (taking into account the propagation loss, slow fading and clutter losses).

$$pl_{it\leftrightarrow vr} = f_{propag}(t_{vr}, h_{vr}, h_{it}, d_{it\leftrightarrow vr}, env) + f_{clutter}(env)$$

where:

h_{vr}: height victim receiver antenna (defined in calculation *dRSS*)

h_{Item}: height interfering transmitter antenna (defined in b))

$d_{it\leftrightarrow vr}$: distance between the affected receiver and the interfering transmitter (defined in b))

$g_{it\to vr}(f_{vr})$: antenna gain of the interfering transmitter in the direction of the affected receiver

$$g_{it\to vr}(f_{vr}) = (g_{it}^{max}, \text{pattern}_{it}) = g_{it}^{max} \times \text{pattern}_{it}(\theta_{it\to vr}, \phi_{it\to vr}, f_{vr})$$

where:

$(\theta_{it\to vr}, \phi_{it\to vr})$: azimuth angles and elevation between the top of the transmitter antenna interferer more próximo and the top of the victim receiver antenna (defined in b))

$g_{vr\to it}(f_{vr})$: antenna gain of the affected transmitter in direction of interferer

$g_{vr\to it}(f_{vr}) = (g_{vr}^{max}, \text{pattern}_{vr}) = g_{vr}^{max} \times \text{pattern}_{vr}(\theta_{vr\to it} + \pi, -\phi_{vr\to it}, f_{vr})$

d) calculation of *iRSSintermod*

$$iRSS_{intermod} = \left(p_{it,k}^{supplied}, g_{it,k}^{PC}, g_{it,k\to vr}, pl_{it,k<\to vr,a_{vr}}, g_{vr\to it,k,sensvr,intermod}\right)$$

where:

$i_{i,j}RSS_{intermod}$: product third order intermodulation of the frequency f 0

$i_{i,j}RSS_{intermod} = 2i_iRSS_{int} + i_jRSS_{int} - 3intermod - 3sens_{vr} - 9 \quad \text{dB}$

Interferer i transmits frequency $f_{it,i} = f_{it}$ and interferer frequency ja $f_{it,j}$ (See b)), which defines $\Delta f = f_{it,j} - f_{it}$ and is obtained $f_0 = f_{it} - \Delta f = 2f_{it} - f_{it,j}$. Assuming an ideal filter (cutting factor 0), the intermodulation product should only be taken into account for bandwidth b:

$f_{vr} - b/2 \le f_0 \le f_{vr} + b/2$

For other cases you can be neglected intermodulation product.

i_kRSS_{int}: power received at the victim receiver due to interferer $k \bullet ia$ fit or interferer $k \bullet fit$

$i_kRSS_{int} = p_{it,k}^{supplied}, g_{it,k}^{pc}, g_{it,k\to vr}, pl_{it,k\leftrightarrow vr}, g_{vr\to it,k}$

The various parameters defined in the above a) to c). To calculate i_iRSS_{int} can use the same algorithms listed in Section 6 because i_iRSS_{int} corresponds to $i_iRSS_{block} + a_{vr}(f_{it}, f_{vr})$.

intermod: intermodulation response to receiving a desired signal is 3 dB above the sensitivity.

Two cases can be foreseen:

Case 1: *intermod*It is given by the user, with typical values of, for example, 70 dB for base station equipment and 65 dB for mobile and portable handheld devices. It is used to calculate generic limits.

case 2: *intermod*(• f) is measured as a function of referring to fvr • f (see Section 9)

PROPAGATION MODEL

The tool several propagation models that depend on the chosen environment for the different cases are offered:

- General environment: open, suburban or urban area;
- Interferers environment: indoors or outdoors;
- Receiver affected environment: indoors or outdoors.

Table 1 shows the domain of validity of the models described:

TABLE 1

Below 30 MHz	No model available. The curves of Recommendation ITU-R P.368 are suitable for high-power transmitters and large distances, so not suitable for interference calculations
Above 30 MHz	Model space: $L[dB] = 32{,}5 + 20\log(f[MHz]) + 20\log(d[km])$ A mandatory condition for use is a clear line of sight; ie the first Fresnel zone must be clear
Between 30 MHz and 3 GHz	Modified model, available for loss calculations outer-outer path Hata. Care should be taken when it is expected that the propagation distances exceeding 20 km. Also suitable are inside-inside and InteriorExterior models For broadcasting the propagation model in Recommendation ITU-R P.1546 applies
Above 3 GHz	It is not recommended to use the modified Hata model. The model spherical diffraction is suitable for an environment of open area and point to point. No model available for suburban and urban environment. Also suitable are inside-inside and InteriorExterior models

For improved flexibility of the tool, the user can also use a «generic» model, for both the desired signal path as the path interference d. The tool user must then enter the parameters A, B and C of the formula on the median value of attenuation and distribution of the variation in path loss Dv. As default distribution, log normal distribution with standard deviation determined by the user is proposed. Then we have:

$$f_{propag}(d) = L + T(D_v)$$

1. Path loss in free space

The path loss in free space are defined as follows:

$$L[dB] = 32{,}5 + 20\log(f[MHz]) + 20\log(d[km])$$

The user can also create and implement more complex models using a simple program. As an example of notation in the propagation model user-defined path loss in free space considering the difference in height of the antennas they are given by:

$L1 = 32.5;$

$L2 = 20 * log10\ (freq\ ());$

$L3 = 10 * log10\ (dist\ () * dist\ () + (hrx\ () * hrx\ () + htx\ () * htx\ ())\ /1000000);$

$L = L1 + L2 + L3;$

2. Hata model modified

$$f_{propag}(f, h_1, h_2, d, env) = L + T(G(\sigma))$$

where:

L: median value of propagation loss (dB)

- standard deviation of the distribution of slow fading (dB)
- *F*: Frequency (MHz)
- H_m: min $\{h_1$, H2$\}$
- *hb*: max $\{h_1$, H2$\}$
- *d*: distance (km), preferably less than 100 km
- *env*: (Outside / outdoor), (rural, urban or suburban), (propagation above or below roof).

If Hm and / or Hb is less than 1 m, you use the value of 1 m. The heights exceeding 200 m antenna can also cause significant errors. The spread under the roof means both Hm and Hb are higher than the height of the roof. In other cases, the spread is above the roof (Hb exceeds the height of the roof).

Calculation of the median value of path loss *L*

Case 1: $d = 0,04$ km

$$L = 32,4 + 20\log(f) + 10\log\left(d^2 + (H_b - H_m)^2 / 10^6\right)$$

case 2: d ³ 0.1 km

$$a(H_m) = (1,1\log(f) - 0,7)\,\text{mín}\{10, H_m\} - (1,56\log(f) - 0,8) + \text{máx}\{0, 20\log(H_m/10)\}$$

$$b(H_b) = \text{mín}\{0, 20\log(H_b/30)\}$$

Note that for SRDs in the case of a low value of the antenna height of the base station, Hb, $b(H_b) = \text{mín}\{0, 20\log(H_b/30)\}$ It is replaced by:

$$b(H_b) = (1,1\log(f) - 0,7)\,\text{mín}\{10, H_b\} - (1,56\log(f) - 0,8) + \text{máx}\{0, 20\log(H_b/10)\}$$

$$\alpha = \begin{cases} 1 \\ 1 + \left(0,14 + 1,87\times10^{-4}\, f + 1,07\times10^{-3}\, H_b\right)\left(\log\dfrac{d}{20}\right)^{0,8} \end{cases}$$

para $d \le 20$ km

para $20 \text{ km} < d \le 100$ km

subcase 1: Urban

30 MHz f= 150 MHz

$$L = 69,6 + 26,2\log(150) - 20\log(150/f) - 13,82\log(\text{máx}\{30, H_b\}) + \left[44,9 - 6,55\log(\text{máx}\{30, H_b\})\right]\log(d)^{á} - a(H_m) - b(H_b)$$

150 MHz f= 1500 MHz

$$L = 69,6 + 26,2\log(f) - 13,82\log(\text{máx}\{30, H_b\}) + \left[44,9 - 6,55\log(\text{máx}\{30, H_b\})\right]\log(d)^{á} - a(H_m) - b(H_b)$$

1500 MHz f= 2 000 MHz

$$L = 46,3 + 33,9\log(f) - 13,82\log(\text{máx}\{30, H_b\}) + \left[44,9 - 6,55\log(\text{máx}\{30, H_b\})\right]\log(d)^{á} - a(H_m) - b(H_b)$$

2 000 MHz f= 3000 MHz

$$L = 46{,}3 + 33{,}9\log(2000) + 10\log(f/2000) - 13{,}82\log(\text{máx}\{30, H_b\}) + \left[44{,}9 - 6{,}55\log(\text{máx}\{30, H_b\})\right]\log(d)^{á} - a(H_m) - b(H_b)$$

subcase 2: Suburban

$$L = L(\text{urban}) - 2\left\{\log\left[(\text{mín}\{\text{máx}\{150, f\}, 2000\})/28\right]\right\}^2 - 5{,}4$$

subcase 3: open area

$$L = L(\text{urban}) - 4{,}78\left\{\log\left[\text{mín}\{\text{máx}\{150, f\}, 2000\}\right]\right\}^2 + 18{,}33\log\left[\text{mín}\{\text{máx}\{150, f\}, 2000\}\right] - 40{,}94$$

case 3:

$$L = L(0{,}04) + \frac{\left[\log(d) - \log(0{,}04)\right]}{\left[\log(0{,}1) - \log(0{,}04)\right]}\left[L(0{,}1) - L(0{,}04)\right]$$

When L is below the free space attenuation for the same distance, you use the value of the free space attenuation.

Estimated standard deviation for lognormal distribution

Case 1: $d = 0{,}04$ km

case 2:

$$\sigma = 3{,}5 + \frac{(12 - 3{,}5)}{(0{,}1 - 0{,}04)}(d - 0{,}04)$$ dB to spread over the roofs

$$\sigma = 3{,}5 + \frac{(17 - 3{,}5)}{(0{,}1 - 0{,}04)}(d - 0{,}04)$$ dB to propagation below the roofs

case 3:

$$\sigma = 12 + \frac{(9 - 12)}{(0{,}6 - 0{,}2)}(d - 0{,}2)$$ dB to spread over the roofs

$$\sigma = 17 + \frac{(9 - 17)}{(0{,}6 - 0{,}2)}(d - 0{,}2)$$ dB for propagation under dand the roofs

3. Spherical diffraction pattern

The spherical spreading model is based on several ITU-R P.452, P.676 and ITU-R ITU-R Recommendations P.5261.

According to ITU-R Recommendation P.452, the median value of loss between transmitter and receiver is given by the equation:

$$L_{bd}(p) = 92,5 + 20\log f + 20\log d + L_d(p) + A_g$$

where:

$L_{bd}(p)$: basic loss (dB), depending on the percentage of time, *p* (%)

F: frequency (GHz)

d: distance (km)

$L_d(p)$: Difra lossesction (dB), depending on the percentage of time p (%)

Ag: attenuation gases and water from the atmosphere (dB).

The atmosphere attenuation is given by:

$$A_g = \left[\gamma_o(f) + \gamma_w(\rho, f)\right] d$$

where:

$\gamma_o(f)$: linear attenuation due to aire dry (oxygen) (dB / km)

$\gamma_w(\rho, f)$: linear attenuation (dB / km) due to water as a function of water concentration

According to Recommendation ITU-R P.676, both terms can be approximated by the equations:

Attenuation due to water:

$$\gamma_w(\rho, f) = \left[0,050 + 0,0021\rho + \frac{3,6}{(f - 22,2)^2 + 8,5} + \frac{10,6}{(f - 183,3)^2 + 9} + \frac{8,9}{(f - 325,4)^2 + 26,3}\right] f^2 \rho \times 10^{-4}$$

for f <350 GHz

Attenuation oxygen:

$$\gamma_o(f) = \left[7,19 \times 10^{-3} + \frac{6,09}{f^2 + 0,227} + \frac{4,81}{(f - 57)^2 + 1,50}\right] f^2 \times 10^{-3} \qquad F = 57\text{ GHz}$$

$$\gamma_o(f) = 10,5 + 1,5\,(f - 57) \qquad F = 60\text{ GHz}$$

$$\gamma_o(f) = 15 - 1,2\,(f - 60) \qquad F = 63\text{ GHz}$$

$$\gamma_o(f) = \left[3,79 \times 10^{-7} f + \frac{0,265}{(f - 63)^2 + 1,59} + \frac{0,028}{(f - 118)^2 + 1,47}\right](f + 198)^2 \times 10^{-3} \quad \text{for} \quad F > 63\text{ GHz}$$

Note that for simplicity has been performed a linear interpolation between 57 and 63 GHz. The maximum is 15 dB / km to 60 GHz.

According to ITU-R Recommendation P.526, diffraction loss Ld (p) can be obtained from the received field strength E with reference to the free space E0:

$$-L_d(p) = 20 \log \frac{E}{E_0} = F(X) + G(Y_1) + G(Y_2)$$

where:

X: normalized radio path between the transmitter and receiver

Y_1: normalized antenna height of the transmitter

Y_2: normal antenna heightlifting the receiver

$$X = 2,2\,\beta\, f^{1/3}\, a_e^{-2/3}\, d$$

$$Y = 9,6 \times 10^{-3}\, \beta\, f^{2/3}\, a_e^{-1/3}\, h_i$$

where:

: parameter obtained from the soil admittance factor K: • 1 for f < 20 MHz

F: Frequency (MHz)

ae: fictional Earth's radius (km) (see definition below)

d: distance (km)

hi: antenna height above ground (m) with i • 1 or 2 for the transmitter or receiver respectively.

The term dependent on the distance F (X) is given by the following semi-empirical formula:

$$F(X) = 11 + 10 \log(X) - 17,6X$$

The gain G antenna height (Y) is given by the following set of formulas:

$$G(Y) = 17,6(Y - 1,1)^{1/2} - 5 \log(Y - 1,1) - 8 \quad \text{for} \quad Y < 2$$

$$G(Y) = 20 \log(Y + 0,1Y^3) \quad 10K < Y < 2$$

$$G(Y) = 2 + 20 \log K + 9 \log(Y/K)\left[\log(Y/K) + 1\right] \quad \text{for} \quad K/10 < Y < 10K$$

$$G(Y) = 2 + 20 \log K \quad \text{for} \quad Y < K/10$$

where:

K: factor normalized admittance of the surface of the Earth (refer to Recommendation ITU-R P.526). Default: 10-5.

Note the different units used for frequency. This variation of the path loss is fictitious variability Earth radius (km), which is considered dependent on the percentage of time p:

$$a_e(p) = 6\,375\ k(p)$$

with fictional coefficient k Earth radius (p) expressed as:

$$k(p) = k_{50} + (5 - k_{50})\frac{(1,7 - \log p)}{(1,7 - \log \beta_0)} \qquad \text{for } p < 50\%$$

$$k(p) = k_{50} \quad \text{for } p > 50\%$$

Y

$$k_{50} = \frac{157}{157 - \Delta N}$$

where:

- average gradient refraction profile radiowave over a layer of 1 km of the atmosphere from the surface. The default is 40 units / km for Europe (normal atmosphere). This value is obtained: k50= 4/3 and ae = 8 500 km.

  NOTE 1 - The average gradient is positive.
- existence probability (%) of super-refraction layer (N < 100 units / km) in the lower atmosphere. Default: 1% for Europe.

The probabilities p 0 are expressed in percent; that is, a range of variation of 0 to 100%.

Normally a value is chosen by default constant $p = 50\%.$ Small percentages of time allow the simulation of anomalous propagation conditions.

To apply this model must take into account the following restrictions:

- The frequency range must be greater than 3 GHz. With due caution can be used lower frequencies, but not below 300 MHz because of the effects of the surface admittance and polarization.

- The model was developed for open areas (rural). Therefore, further attenuation obstacles such as buildings present in an insuburban or urban lathe is not included.
- They have not been taken into consideration losses rain.
- This model applies only to terrestrial radio paths.

PROPAGATION MODELS COMBINED INSIDEEXTERIOR

Most propagation models published in the literature in the public domain are calculated for an application interior or exterior. Actually a combination of both types is needed, in the «real world.»

For mixed cases have combined classical models outside the Hata and the model spherical diffraction (ITU-R Recommendations P.452, P.526 and ITU-R ITU-R P 0.676), with an interior model. An illustrative description appears. The path loss pL are composed of the median value of the path loss L and the Gaussian variation T (G (•)), the standard deviation being:

$$p_L\left(f, h1, h2, d, env\right) = L + T\left(G\left(\sigma\right)\right)$$

where:

F: Frequency (MHz)

h_1: height of the transmitter antenna (m)

h_2: antenna height of the receiver (m)

d: distance (km)

env: parameter representative environments transmitter and receiver.

For exteriorinterior is met:

Location: the transmitter and receiver are both on the outside

Modified Hata model:

Median value: L (Outdoor - outdoor) • LHata (outdoor - outdoor)

Variation: intrinsic variation, •outdoor - outdoor)

Spherical diffraction pattern

Median value: L (Outdoor - outdoor) • Lspherical

Case 1: InteriorExterior or exteriorinterior

- Location: the transmitter is low ceiling and the receiver on the outside, or vice versa
- Modified Hata model:

Median value: L (indoor - outdoor) • LHata (outdoor - outdoor) Lwe

where Lwe is the attenuation due to external walls (default • 10 dB).

Variation: • (indoor - outdoor)

$$\sqrt{\sigma_{Hata}^2 + \sigma_{add}^2}$$

- being •add additional standard deviation of the signal (default • 5 dB).

It increases the standard deviation of lognormal distribution compared to exteriorexterior situation, because of the additional uncertainty as to the materials and the relative location of the building.

Spherical diffraction pattern

Median value: L (indoor - outdoor) • Lspherical Lwe

Variation: • (indoor - outdoor) = •add

The log-normal distribution is determined by the additional variation due to differences among building materials only as to the spherical diffraction pattern is not taken into account some variation.

case 2: Interior Exterior

There are two different scenarios possible: The transmitter and receiver are in the same building or in two different buildings. It randomly chooses one of the two situations.

a) Selection of the situation

The first step is to determine whether or not interiorinterior situation corresponds to the case of a transmitter and a receiver located in the same building. This is done by calculating the random variable SB (same building, same building).

b) InsideInside, various buildings

- Location: transmitter and receiver in different buildings: *P*(Yes) • 0 or P (No) • 1
- Modified Hata model:

Median value: L (indoor - indoor) • LHata (outdoor - outdoor) 2LWE

It is noteworthy that must be added the losses due to 2 outdoor walls.

Variation: •(indoor - indoor) $\sqrt{\sigma_{Hata}^2 + \sigma_{add}^2}$

- Spherical diffraction pattern

Median value: L (indoor - indoor)

Variation: • (indoor - indoor) $= \sqrt{2}\sigma_{\text{add}}$

The log-normal distribution is determined by the additional variation due to differences among building materials only as to the spherical diffraction pattern is not taken into account some variation. The variation increases by the second outer wall.

c) InsideInside, the same building

- Location: transmitter and receiver in the same building: *P*(Yes) • 1 or P (No) • 0
- Model propagation interior:

median value: $L(Indoor)$=-27.6+20log(1000d)+20log(f)+fix($\frac{1000d}{d_{room}}$) $L_{wt} + k_f^{\left[\frac{k+2}{kj+1}-b\right]} Lf$

with:

$$k_f = fix\left(\frac{|h2-h1|}{h_{floor}}\right)$$

Lwi: the inner wall loss (dB)

Lf: loss between adjacent floors (dB)

b: paramempirical eter

droom: Room size (m)

hfloor: each floor height (m)

Note that the path length d It is expressed in km, while for the drive frequency is MHz.

Variation: • (indoor - indoor)

The test lognormal distribution is performed using a standard deviation by the user and covering the variation within the building due to its design, furniture rooms, etc. The default is •in = 10 dB.

PROPAGATION MODEL FOR VHF AND UHF (ITUR P.1546)

Propagation curves obtained for broadcasting in Recommendation ITU-R P.1546 based on the previous Recommendation ITU-R P.370: A set of received field strengths E (dB (μV / m)) normalized to a transmitter power of 1 kW pra Using the conversion shown in rcommendation P.525 UITR. This field strength level can become the median value of the basic radio path loss L (dB) between two isotropic antennas by the following equation: $L(p_1, p_t) = 139.4 + 20logf[MHz] - E(f, d, h_1, env)$

being

pl: 50% of the darker corners ofyou

pt: 50, 10, 5 or 1% of the time

F: within the range 30-250 MHz and 000 MHz 450-1

d: distances between 10 and 1 000 km

h_1: antenna height variable transmitter of between 37.5 and 1200 m

env: different types of environments: land (used in SEAMCAT), hot or cold sea.

Note that the path loss should not be less than the path loss in free space.

Path loss, pl, including variation sites can be expressed as the sum of the median values of path loss and a Gaussian distribution:

$$pl = L(p_t, p_l = 50\%) + T(G(\sigma))$$

ITU-R Recommendation P.1546 proposes a propagation model for a prediction point zone field strength, mainly for broadcasting but also for land mobile services and maritime mobile and some fixed (for example, those using point to multipoint) systems in the frequency band 30 MHz and 3 000 for a range of distances of 1 km 1 000 km. For analyzing compatibility scenarios may assume the following simplifications:

- level ground;
- restricting the spread on land only; ie exclusion of mixed and sea paths;
- positive antenna heights, just twentyand.

The parameters of this model are listed propagation:

to) Path dependent parameters (constant during a simulation for a specific distance) are:

- percentage of time: *pt* (%);
- analog / di: transmitter systemgital;
- bandwidth transmitter: *bt* ;
- global environment: rural, suburban, urban.

b) variable parameters (vary for each event simulation):

- height of the transmitter antenna: *ht* (M);
- antenna height of the receiver: *hr* (M);
- Frequency f (MHz);
- distance *d* (Km).

To calculate the path loss according to Recommendation ITU-R P.370 the following procedure should apply:

Step 1: Verification of the application range of propagation model regarding PERCENTAGEAJE time, frequency, distance and antenna height:

- Percentage of time: 1% < *pt* <50%, for pt ³ 50% pt is set to a value = 50%;
- Frequency: 30 MHz < *F* <3 000 MHz;
- Distance: 0,001 km < *d* <1 000 km;
- Antenna height transmisor: 0 m <ht <3000 m;
- Height of the receiver antenna 1 m < *hr* <3000 m

Step 2: Determining upper and lower nominal percentages *ptinf* and pt-sup:

If pt <10 then ptinf ptsup = 1% = 10%, for other ptinf values = 10% and 50% ptsup =

Step 3: Determining upper and lower nominal frequencies:

If f <600 MHz = 100 MHz then finf and FSUP = 600 MHz

for other values finf = 600 MHz and 000 MHz FSUP = 2

step 4: Yes *ht*³ 10 m: the field strength E = (f = f, d, ht, hr, pt) is calculated according to the substeps 4.1-4.4

4.1 substep: Calculation of the four following field strengths:

- *AND*(F • finf, d, ht, hr, ptinf)
- *AND*(F • FSUP, d, ht, hr, ptinf)
- *AND*(F • finf, d, ht, hr, ptsup)
- *AND*(F • FSUP, d, ht, hr, ptsup)

according to the procedure described in substeps 4.1.1. to 4.1.4.

4.1.1 substep: the value of the dimensionless parameter is determined *k*Which it is a function of the height of the transmitter necessary, ht, as follows:

$$k = \frac{\log\left[\dfrac{h_t}{9{,}375}\right]}{\log(2)}$$

4.1.2 substep: It is determined from the following table parameter set to_0 to a3, b0 to b7, c0 and d0 to d1 c6 to be used according to the values of the frequencies and time percentages:

Frequency	100 MHz			600 MHz			2000 MHz		
pt (%)	fifty	10	1	fifty	10	1	fifty	10	1
to_0	0.0814	0.0814	0.0776	0.0946	.0913	0.0870	0.0946	0.0941	0.0918
to_1	0.761	0.761	0.726	.8849	.8539	.8141	.8849	.8805	.8584
to_2	-30.444	-30.444	-29.028	-35.399	-34.160	-32.567	-35.399	-35.222	-34.337
to_3	90.226	90.226	90.226	92.778	92.778	92.778	94.493	94.493	94.493
b_0	33.6238	40.4554	45.577	51.6386	35.3453	36.8836	30.0051	25.0641	31.3878
b_1	10.8917	12.8206	14.6752	10.9877	15.7595	13.8843	15.4202	22.1011	15.6683
b_2	2.3311	2.2048	2.2333	2.2113	2.2252	2.3469	2.2978	2.3183	2.3941
b_3	.4427	.4761	.5439	.5384	.5285	.5246	.4971	.5636	.5633
b_4	1.256_10-7	7.788_10-7	1.050_10.0 6	4.323_10.0 6	1.704_10-7	5.169_10-7	1.677_10-7	3.126_10-8	1.439_10-7
b_5	1.775	1.68	1.65	1.52	1.76	1.69	1.762	1.86	1.77
b_6	49.39	41.78	38.02	49.52	49.06	46.5	55.21	54.39	49.18
b_7	103.01	94.3	91.77	97.28	98.93	101.59	101.89	101.39	100.39
c_0	5.4419	5.4877	4.7697	6.4701	5.8636	4.7453	6.9657	6.5809	6.0398
c_1	3.7364	2.4673	2.7487	2.9820	3.0122	2.9581	3.6532	3.547	2.5951
c_2	1.9457	1.7566	1.6797	1.7604	1.7335	1.9286	1.7658	1.7750	1.9153
c_3	1.845	1.9104	1.8793	1.7508	1.7452	1.7378	1.6268	1.7321	1.6542
c_4	415.91	510.08	343.24	198.33	216.91	247.68	114.39	219.54	186.67
c_5	.1128	.1622	.2642	.1432	.1690	.1842	.1309	.1704	0.1019
c_6	2.3538	2.1963	1.9549	2.2690	2.1985	2.0873	2.3286	2.1977	2.3954
d_0	10	5.5	3	5	5	8	8	8	8
d_1	-1	1	2	1.2	1.2	0	0	0	0

4.1.3 substep: the maximum field strength is calculated, *Eu*, To the distance d, and for a height ht of the transmitting antenna as follows:

$$E_u = p_b \log\left[\frac{10^{(E1+E2)/pb}}{10^{(E1)/pb} + 10^{(E2)/pb}}\right]$$

being:

$$p_b = d0 + d1\sqrt{k}$$

Y: $E1 = \left(a_0 k^2 + a_1 k + a_2\right)\log(d) + 0.1995k^2 + 1.8671k + a_3$

being:

$$Eref = b_0\left[\exp\left[-b410^{\varepsilon}\right] - 1\right] + b_1 \exp[-\left(\frac{\log(d) - b2}{b3}\right)^2] - b_6 \log(d) + b_7$$

being:

$\xi = \log(d)^{b_5}$

Y:

$$E_{off} = \frac{c_0}{2} \cdot k \cdot \left[1 - tgh \left[c_1 \cdot \left[\log(d) - c_2 - \frac{c_3{}^k}{c_4} \right] \right] \right] + c_5 \cdot k^{c_6}$$

4.1.4 substep: the relationship between the intensity value of the combined field and the free space, Eb, to a distance d to a height ht transmitter is calculated as follows:

$$E_b = p_{bb} \log \left[\frac{10^{(Eu+Ef)/pbb}}{10^{(Eu)/pbb} + 10^{(Ef)/pbb}} \right]$$

being:

efs: the field strength in free space

efs= 106.9 - 20 log (d) dB (μV / m)

pbb: combination coefficient set to a value eight.

4.2 substep: Calculation of the field strength E (f, d, ht, hr, ptinf) using the log-linear interpolation in the frequency range:

AND • $Einf\left(Esup\ Einf\right)log\left(f\ /\ finf\right)/log\left(FSUP\ /\ finf\right)dB\left(V\ /m\right)$

4.3 substep: Double calculating the field strength E (f, d, ht, hr, ptsup) using the log-linear interpolation in the frequency range:

AND • $Einf\left(Esup\ Einf\right)log\left(f\ /\ finf\right)/log\left(FSUP\ /\ finf\right)dB\left(V\ /m\right)$

4.4 substep: Calculation of the field strength E(F, d, ht, hr, pt) using the formula log-linear interpolation in the range of percentage of time:

AND

$$Esup\left(Qinf\ \ Qt\right)/\left(Qinf\ \ Qsup\right)\ Einf\left(Qt\ Qsup\right)/\left(Qinf\ \ Qsup\right)dB\left(mV\ /m\right)$$

where (Qi (x)) is the function inverse cumulative normal distribution complementary:

step 5: For antenna height transmitter / ht basis less than 10 m field strength for the height and distance required is determined using the following method:

The procedure for extrapolating field strength at a required distance d (km) for values of ht in the range of 0 m to 10 m based on distances of the horizon level ground (km) expressed as $d_H(h) = 4,1\sqrt{h},$ where h is the required value of antenna height transmitter / base ht (m).

For d dH (ht) the field strength is given by the curve height of 10 m to its distance from the horizon, •*AND*, being •*AND* the difference in field strengths on the curve height of 10 m at a distance d and a horizon distance ht.

to d ≥dH (ht) the field strength is given by the curve height of 10 m distance•*d* beyond its horizon distance, being •*d* the difference between d and the horizon distance ht.

This can be expressed by the following formulas where E10 (d) is the field strength (dB (microvolts / m) calculated for a transmitter antenna 10 m in height and a distance d (km) according to the procedure described in Step 4:

If in the last equation dH (10) d –dH (ht) exceeds the value of 1000 km, even if d= 1000 km, E10 can be determined from a linear extrapolation for log (distance) of the curve given by:

$$AND \left(Esup\ Einf\right) log\left(d\ /Dinf\right) /log\left(Dsup\ /Dinf\right) dB\left(mV\ /m\right)$$

being:

Dinf: ultimate tabulation distance (km)

Dsup: final tabulation distance (km)

einf:field strength for the penultimate distance tabulacion (dB (microvolts / m))

Esup:field strength at final tabulation distance (dB (microvolts / m)).

NOTE 1-This Recommendation is not valid for distances exceeding 1000 km. This method is only for extrapolation ht <10 m.

step 6: If the antenna height of the receiver / mobile *hr* is not equal to the echo height representative parasite in its location (called R), the value of field strength as follows corrected:

The values of field strength given by the curves of earth and tabs corresponding in this Recommendation refer to an antenna receiver / mobile reference at a height R (m), representing the height of the ground cover surrounding the receiver antenna / mobile, subject to a minimum value of 10 m height. Examples of reference heights may be mentioned 20 m for an urban area, 30 m for a dense urban area and 10 m for a suburban area.

If the receiver antenna height / hr mobile (m) is different from R, a correction must be introduced into the measured field strength from the curve.

If the receiver antenna / mobile is adjacent land should be considered first elevation angle beam arrival calculating a representative echo height modified R '(m), given by:

ht expressed in meters and distance d in km.

If necessary, it is limited value R 'such that not less than 1 m.

When the antenna receiver / mobile is in an urban environment the correction is given by:

$$J(V) = \left[6.9 + 20\log\left(\sqrt{(v-0.1)^2+1} + v - 0.1\right)\right]$$

where:

=

$Hdif = R$ ' - hr m

$$Clut = arctan\left(Hdif\ /15\right) degrees$$

$K_{hr} = 3.2 f)$

$KNU = 0.0108$

F : MHz frequency.

If the antenna receiver / mobile is adjacent to land in a rural environment the correction is given by equation (2) for all values hr.

If the required distance is equal to or greater than the correction d10 again to the required value of h2 should be calculated using equation (2) setting R '10 m.

C_{10}: correction for the required value *hr* the distance d10 using equation (2) and setting R to a value of 10 m

d_{10}: distance at which the path has a Fresnel clearance from 0.6 to *hr* = 10 m calculated as D06 (f, ht, 10) (see Note 2)

DHR: distance at which the path has a Fresnel clearance from 0.6 to the required value hr calculated as D06 (f, ht, hr) (see Note 2).

This Recommendation is not valid for antenna heights receiver / mobile hr, less than 1 m.

step 7: sAdd a normal log-term *G*(•L) for the variability in the percentage of site:

The values of standard deviation for digital systems with a bandwidth less than 1 MHz for analog systems and are expressed in terms of frequency being:

K

- 2.1 for mobile systems in urban locations
- 3.8 for mobile systems in suburban locations or situated between steep hills little
- 5.1 for analogue broadcasting systems

In the case of digital systems with a bandwidth of 1 MHz or higher, it should be used for all frequencies a standard deviation of 5.5 dB.

step 8: If necessary to limit the resulting field strength to the maximum calculated value of the fofollowing signature:

The field strength should not exceed a maximum value Emax given by:

$Emax \quad Efs\, dB(mV/m)$ for land paths

being *efs* field strength in free space for 1kW pra expressed by:

$$efs \bullet 106.9 \;\; 20 log(d) dB(mV/m)$$

step 9: the field strength is converted to losses I path using the following formula:

$$lb = 77.2 - E = 20 log\, f\, dB$$

being:

L_b: basic transmission loss (dB)

AND: field strength (dB (μV / m)) measured with a transmitter power of 1 W EIRP

F: frequency (MHz).

NOTE 1 - The following approximation function inverse cumulative normal distribution complementary, Qi (x) is valid for 0.01 = x = 0.99:

$$Qi(X) = T(x) - (X) \, if \; x = 0.5$$

$$Qi(X) = -\{T(1-x) - (1-x)\} \, if \; x > 0.5$$

being:

$$T(x) = \sqrt{-2\ln(x)}$$

$$\varepsilon(x) = \frac{[(C2T(x) + C1)T(x)] + C0}{[(D3T(x) + D2)T(x) + D1]T(x) + 1}$$

NOTE 2 - The path length that achieves a 0.6 the first Fresnel zone for a smooth curved Earth at a given frequency and antenna heights ht and hr is given approximately by:

$$D_{06} = \frac{D_f D_h}{D_f + D_h} km$$

being:

Df: dependent term frequency

- $0{,}0000389 \, f \, h_1 \, h_2 \; km$

Dh: an asymptotic term defined by horizon distances

- $4{,}1\left(\sqrt{h_t} + \sqrt{h_r}\right) \; km$

F: Frequency (MHz)

ht, Hr: antenna heights above a smooth ground (m).

In the above equations, the value of ht may be limited if necessary so that not less than 0. In addition, they also limited the resulting values of D06 if necessary so as not lower than 0,001 km.

NOTE 3 - the case is not contemplated ht less than zero in Recommendation.

NOTE 4 -no correction due to terrain clearance angle does not apply. Power control function is given by:

$$g_{it}^{PC} = f_{PC}\left(p_{it}^{supplied}, g_{it\to wr}, pl_{it\to wr}, pc_{it}^{t_{hold}}, pc_{it}^{dyc_{rg}}, pc_{it}^{t_{hold}}\right)$$

$$P = f\left(p_{it}^{supplied}, g_{it\to wr}, pl_{it\to wr}, g_{wr\to it}\right) = p_{it}^{supplied} + g_{it\to wr} - pl_{it\to wr} + g_{wr\to it}$$

P: power received by the intended recipient; for example, a base station closed sistsinterfering ma

where $p_{it}^{supplied}, g_{it\to wr}, pl_{it\to wr}$ and $pl_{it\leftrightarrow wr}$ They are defined in paragraphs concerning the calculation of SSRIs. $p_{it}^{t\_hold}$ It is the lowest (minimum) threshold receptor

$Case 1:$

$$p_{it}^{supplied\_PC} = p_{it}^{supplied}$$

$$g_{it}^{PC} = 0$$

$Case(I\ 1):$

$$p_{it}^{supplied\_PC} = p_{it}^{supplied} - (i-1)\,pc_{it}^{st\_rg}$$

$$g_{it}^{PC} = -(i-1)\,pc_{it}^{st\_rg}$$

where i is an integer from 1 to N $steps\,is$

$Case(N\ steps\,2):$

$$p_{it}^{supplied\_PC} = p_{it}^{supplied} - pc_{it}^{dyc_{rg}}$$

$$g_{it}^{PC} = -pc_{it}^{dyc_{rg}}$$

Distribution definitions

- *gaussian distribution* :
- *Rayleigh distribution* :
- User-defined distribution: it should be considered the option of including in herramienta distribution defined by the user.
- discrete distribution:

This is a special distribution with a lower limit *xmin*An upper limit Xmax, and step ^{S} between samples xi. A common example is the distribution of discrete frequencies with a separation of channels constant.

The corresponding distribution *xi* It defined by the following equation:

$$x_i = Xmin + \frac{S}{2} + (i-1)S$$

being :,

In the case of a uniform distribution, each value equal probability $P(xi) = 1/N$ is assigned. In the case of nonuniform distribution, each value to a specific weight factor Pi with the limitation that the sum of these weighting factors must be equal to one is assigned.

Pseudorandom number generation

- From a uniform distribution $U(0,1)$ [Knuth, 1969; Rubinstein, 1981]

$$u_{i+1} = T\left(U(0,1)\right) = \frac{x_{i+1}}{m}$$

where: $x_{i+1} = (ax_j)(mod\, m)$

- *to*: multiplier; for example, 16,807
- *m*: module; for example, *m*• 231-1 • 2 147 483 647
- x_0: initial value, an integer variable having a value between 1

and (m - 1)

- Given a Gaussian distribution $G(\sigma)$

$$T(G(\sigma)) = v1\sqrt{\frac{-2\ln(s)}{s}}$$

where:

for $s\ 1, d =$

v_1 and v2 are two independent random variables (using two different starting values), with uniform distribution between -1 and 1.

- Given a Rayleigh distribution $R(\sigma)$

$$T(R(\sigma)) = \sqrt{(v_1^2 + v_2^2) \times \frac{-2\ln(s)}{s}}$$

where:

for $s\ 1, d =$

v_1 and v2 are two independent random variables (using two different starting values), with uniform distribution between -1 and 1.

From any distribution with a cumulative distribution function given cdf.

Some tests may be performed according to user defined distribution, F.

The book is based on the use of reverse cumulative distribution function, cdf -1, corresponding to the user distribution, F, applied to the result of a sample uniformly between 0 and 1 defined.

where (Uniform test between 0 and 1)

Función de distribución acumulativa directa

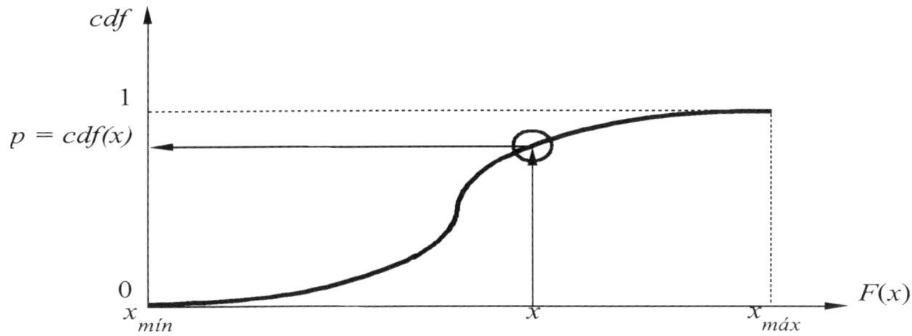

Rap 2028-07

Función de distribución acumulativa inversa

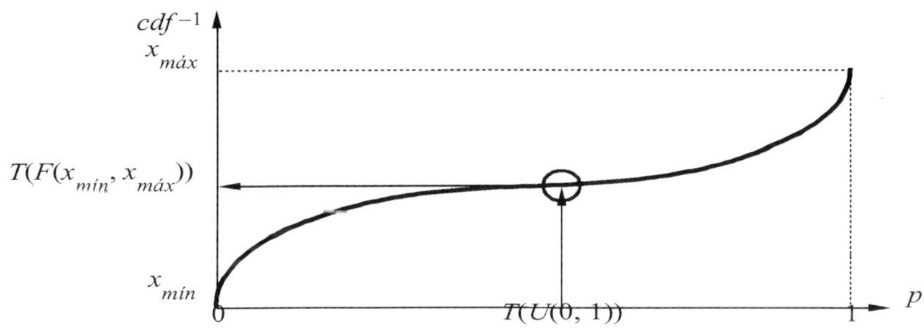

Rap 2028-08

Blockade receptor

1. Basic concepts

The receiver tunes any unwanted signal because the filter is not ideal.

Definition: Blocking measures the ability of the receiver to receive a desired signal and modulated input in the presence of an input signal unwanted frequencies other than those of spurious or adjacent channel responses without these input signals unwanted causing degradation of the higher functioning receiver a specified limit (Document I-ETS 300 113: 1992).

2. Lock level measurement

- Adjust the desired signal level to limit bit error ratio (BER).

- Increasing the level of this desired signal by 3 dB and add the interfering signal, increasing to obtain equal BER.
- The ratio (interfering signal / desired signal) is the value of receptor blockade.

Receptor intermodulation

The main contribution to intermodulation interference comes from the interfering signals in neighboring channels due to the frequency selectivity of antennas and receiving equipment. a service with a desired frequency signal f0 is supposed, a separation channel•*F* and interfering signals at frequencies Ei1 and Ei2 $f_0 + n\Delta f$ Y $f_0 + 2n\Delta f$ respectively. The nonlinearity of the receiver generates intermodulation product Eif third-order frequency

$$f_0 = 2(f_0 + n\Delta f) - (f_0 + 2n\Delta f),\ n = 1, 2, \ldots \tag{3}$$

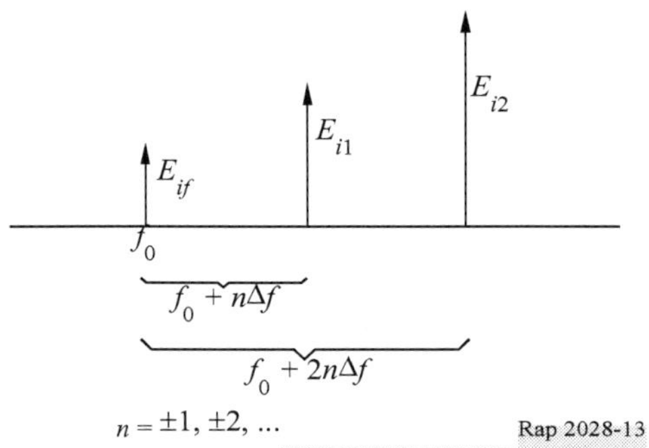

Eif intensity intermodulation product signal is given by the expression:

$$E_{if} = kE_{i1}^2 E_{i2} \tag{4}$$

with constant ka determined. For signal levels (in dB) Equation (4) takes the form:

$$L_{if} = 2L_{i1} + L_{i2} + 20\log k \tag{5}$$

20 log k constant of equation (5) can be determined from the measurement procedure described in the ETS 300-113 standard, § 8.8 of the European Institute of Telecommunications Standards (ETSI). The method is similar to the contribution in Section 7 for blocking interference.

The ETS 300-113 defined by the response to LIMR intermodulation interference signal levels Li1 = Li2 from which start recording bit errors due to intermodulation (see Fig. 14).

This means that for Li1 and Li2 as in Fig. 14, it has an intermodulation product Lif whose value is exactly the noise floor (0 dB). . Substituting Li1 and Li2 Fig 14 in equation (5) is obtained:

$$0 = 2\left(L_{imr} + 2dB + L_{sens}\right) + \left(L_{imr} + 3dB + L_{sens}\right) + 20logk \quad (6)$$

With the value of k obtained in equation (6), equation (5) becomes:

$$L_{if} = 2L_{i1} + L_{i2} - 3L_{imr} - 3L_{sens} - 9dB \quad (7)$$

Influence of different bandwidths

a. desired path

The desired transmitter transmits its pwt power (dBm) RVF frequency within a width determined, bvr band. This bandwidth is also used to determine the intermodulation products.

b). interferer

For the interfering transmitter, you define a template emissionit emission as a function of f - fit as the maximum power levels emissionit (•f) in the reference bandwidth bs (• f) as specified by the user. This template can also be expressed as the maximum value of:

- The sum of the interference power $p_{it}^{supplied}$, a template relative emission (containing the desired transmission and all unwanted emissions, including emission threshold depending on the power control) and the gain of power control;
- Or the absolute emission threshold.

The relative emission template is described by a triplet of parameters (frequency drift (MHz), relative emission level (dBc) and reference bandwidth (MHz)). The minimum emission threshold defined in paragraph e) of this Section.

Power interferer pit (dBm) to fit is used to evaluate the link budget with the desired receiver (i.e., power control).

c. Principles of determining the interference power

Interfering frequencies fall exactly on the side of the receiver affected (cochannel interference).

To simplify algorithms, pmi template function is normalized to the reference bandwidth of 1 Hz:

$$p_{ni} = p_{mi}(\Delta f) - 10\log\frac{b}{1Hz}$$

Bandwidth b It is the template used for emission.

The total received power interfering emissionit can easily be calculated by integrating over the bandwidth of the receiver from a • fvr - fit - bvr / 2 to b = fvr - fit bvr / 2

$$power_{it} = 10\log\left\{\int_a^b 10 \wedge \left(\frac{p_{n_{it}}(\Delta f)}{10}\right) d\Delta f\right\}$$

where PNI is the standard template (dBm / Hz). Using a reference bandwidth of 1 Hz, the integral can be replaced by a summation where Powerit is given in dBm:

$$power_{it} = 10\log\left\{\sum_{i=a}^{b} 10 \wedge \left(\frac{p_{n_{it}}(\Delta fi)}{10}\right)\right\}$$

NOTE 1 - interference power of a radio system has another bandwidth can be estimated using the above algorithms. This calculation is only necessary for the interference due to unwanted emissions, or cochannel, but not for blocking and intermodulation.

It should be noted that should always apply a template user-defined, even when the template is flat.

d. Calculation of emissions

The principle is explained in c). However, these algorithms are very slow so it refers to computing time. Consequently, the problem is addressed as follows:

The total interference power on the carrier emission_relit, can be calculated by integrating over the bandwidth of the receiver between a • $fvr - fit - bvr2 to b = fvr - fit = bvr \ 2$

$$emmision_{rel_{it}} = 10\log\left\{\int_a^b P_{rel}^{linear}(\Delta f)\,d\Delta f\right\} = 10\log\left\{\int_a^b 10^{\frac{P_{rel}^{dBc}(\Delta f)}{10}}\,d\Delta f\right\}$$

where P_{rel}^{dBc} It is the standard template by the user (dBc / Hz) defined.

The template is expressed as a matrix of N 1 points $(\Delta f_i, P_i)$ and it is assumed to be linear between these points.

$$P_{rel}(\Delta f) = P_i + \frac{\Delta f - \Delta fi}{\Delta f_{i+1} - \Delta f}(P_{i+1} - P_i)$$

This leads to:

$$emmision_{rel_{it}} = 10\log\left\{\sum_{i=0}^{N-1}\int_{\Delta f_i}^{\Delta f_{i+1}} 10^{\frac{P_{rel}^{dBc}(\Delta f)}{10}}\,d\Delta f\right\}$$

where:

$$\Delta f_0 = a = f_{vr} - f_{it} - B_{vr}/2$$

$$\Delta f_N = b = f_{vr} - f_{it} + B_{vr}/2$$

The following equation is obtained:

$$emmision_{rel_{it}} = 10\log\left\{\frac{10}{ln10}\sum_{i=0}^{N-1}\frac{\left(P_{i+1}^{linear} - P_i^{linear}\right)\left(\Delta f_{i+1} - \Delta f_i\right)}{\left(P_{i+1}^{dBc} - P_i^{dBc}\right)}\right\}$$

e. minimum threshold for unwanted emission

The formulas mentioned above also apply to the absolute minimum threshold emission floorit emission_ (dBm). Template emission threshold can be described by a triplet of parameters (frequency drift (MHz), reference bandwidth (MHz), and minimum emission threshold (dBm)).

The actual emission is limited by the minimum emission threshold by the following expression:

$$emmision_{it} = \max\left(emmision_{rel_{it}} + p_{it}^{supplied} + g_{it}^{PC}, emmision_{floor_{it}}\right)$$

Note that the comparison involves the gain of power control if selected power control.

It should also be noted that the minimum threshold SEAMCAT unwanted emission means 1 MHz.

Radio cell size in a noise limited network

Assuming that the received power is equal to the victim receiver sensitivity, the radius Rmax for the desired radio path can be determined by the following equation:

$$f_{median}\left(f_{vr}, h_{vr}, h_{wt}, R_{max,} env\right) + f_{slowfading}\left(X\%\right) = P_{wt} + g_{wt} + g_{vr} - sens_{vr}$$

where the path loss are defined by a median value of the loss plus an additional term representing the distribution

$$ploss = f_{median} + f_{slowfading}\left(X\%\right)$$

Distribution losses Ploss path can generally be expressed by the following equation:

$$Q\left(\mu + a, R_{max}\right) = y$$

where Q It is the cumulative distribution for Rmax and losses resulting average path m and additional path loss due to availability or coverage y. Loss of coverage x It corresponds to y by the relation 1 - y. Assuming that the slow fading can be approximated by a lognormal distribution, ie median value $\approx$ average value, it may be introduced to the relationship where • b• b It is a multiple of the standard deviation • known. Here are some illustrative examples appear: for 95% coverage, b takes a value of 1.96; 99% for a value of 2.58; 99.9% to a value of 3.29; b • 1 for coverage is 68%; b • 2 to 95.5%. The exact values can be readily determined using the inverse Gaussian function.

Therefore the transcendental equation:

$$g\left(R_{max}\right) = P_{wt} + g_{wt} + g_{vr} - sens_{vr} - f_{median}\left(f_{vr}, h_{vr}, h_{wt}, R_{max,} env\right) - b\sigma$$

It can be solved using a linear iteration by counterfeiters regular method:

$$\tilde{R}_{max} = R_{max0} - \frac{R_{max0} - R_{max1}}{g\left(R_{max0}\right) - g\left(R_{max1}\right)} g\left(R_{max0}\right)$$

Note that can be obtained faster convergence distance by applying the logarithmic scale; ie the variable R should be replaced by log (R).

Note that in this case must be reversed formulas given for $f_{median}\left(R_{max}^{wt}\right)+\ldots$

There are three distinct to describe the antenna pattern forms:

- omnidirectional antenna;
- directive antenna pattern (dBi); the gain refers to the main lobe and depends on the azimuth angle and elevation angle;
- symmetric antenna diagram.

These diagrams are often used by fixed and space services. According to Recommendation ITU-R IS.847 or UITR F.699 antenna gain g (DBi) can be expressed by the following equation:

$$g = g_0 - 10 \log(D/\lambda) - 25 \log \psi \quad for$$

being

g0: the maximum gain of the main lobe (dBi); for example, 52 dBi

D: the diameter of the parabola antenna (m)

λ: wavelength 300 / f [MHz] (m)

ψ: spherical angle (degrees) formed by the direction considered the main beam, defined for

ψ 0: limit for the main lobe (degrees); for example 48 °.

It is not allowed to use the antenna for spherical angles between the direction considered and main beam axis (to be defined in an antenna in elevation and azimuth antenna diagram). The spherical angle ψ, symmetrical with respect to its axis, is a combination of elevation and azimuth angles according to the following equation:

$$\cos(\omega) = \cos(\theta)\cos(\varphi)$$

being

θ: the azimuth angle (degrees)

φ: the elevation angle (degrees)

The gain outside the main lobe should be defined as a fixed value that covers the entire range of angles.

Evaluation of device distribution

The DEE flowchart can test for goodness of fit for the chi-square method or the Kolmogorov-Smirnov algorithm (used in SEAMCAT).

Basically, the algorithm checks if a random sample of observations meets a cumulative distribution prespecified. The predefined distribution can be continuous, discrete or hybrid. Thus, the chi-square method is extremely versatile and unique algorithm for use in DEE order to test all possible types of probability distribution functions is proposed.

It is introduced into the DEE a matrix of samples of the random variable RSS. The DEE first checks if the matrix has sufficient length N (number of samples) to generate a stable distribution. This is achieved using N - dN samples to find a function of initial discrete distribution and calculate the cumulative distribution function corresponding, which is then used as reference in the chi-square test applied to the total population of N samples. If the test shows that two discrete distributions differ by an amount greater than an acceptable and prespecified value, a message is sent to EGE to generate some additional samples. Conversely, if the chi square criterion is satisfied, the DEE go ahead with the test.

The flowchart, for example, shows a Gaussian distribution test. Chi squared algorithm is equally applicable to any other continuous distribution that could be representative of RSS random variable. A continuous distribution function allows a closed form expression for calculating the probability that the ICE, in turn, ensures efficient numerical calculation. If no function continuous probability distribution that fits the sample population with adequate accuracy, the only way to proceed is to use discrete probability distribution function and perform numerical calculation of probability.

The notation used is:

- *N*: size of the sample population
- *I*: internal counter for stability testing results
- *dN*: portion of the population size (eg dB • 0.1N)
- *Y*: criteria chi square test (see Section 1 to Annex 3)

$\chi_{1-\alpha}$: quantile - reference level for chi square test

n: total counter shows

TEST CHI SQUARE GOODNESS OF FIT

Test goodness of fit chi square test is one of the oldest and best known statistics.

Let X1, X2,. . . XN a sample taken from a population of unknown function cumulative distribution (cdf) Fx (x). The test is based on chi square test the null hypothesis:

H0Fx (x) • F0 (x) for all x as the alternative H1: Fx (x)' F0 (x) for some values of x

Assume that N observations are grouped in K mutually exclusive categories. Nj is the observed number of tests in the jth category (j • 1, 2, ..., K). In addition, it N_j^0 the number of tests is expected to fall in the jth category according to the known cdf, F0 (x).

The real test uses the following criteria:

$$Y = \sum_{j=1}^{K} \frac{\left(N_j - N_j^0\right)}{N_j^0}, \sum_{j=1}^{K} N_j = N$$

it takes a small value when H0 is true and large when H0 is false. Y is a random variable which obeys chi-square distribution for one value large N.

In practice, that the hypothesis H0 compliance must be satisfied that

$$P\left(Y > \chi_{1-\alpha}^2\right) = \alpha$$

where αis the significance level, for example 0.05 or 0.1; quantile $\chi_{1-\alpha}^2$ Which corresponds to the probability of 1 -aIt appears in the tables of the chi-squared distribution.

The goodness of fit test chi square applies equally density functions discrete and continuous probability.

Stability test Kolmogorov-Smirnov

The purpose of this evaluation stage is the estimate if the number of generated events is enough to consider the stable results from a statistical point of view. The stability assessment is performed by means of a test of goodness of fit with the Kolmogorov-Smirnov to check whether the distribution obtained with N - dN samples and obtained with N samples do not differ by more than a specified value.

First, they must obtain two cumulative distribution functions from the matrix input vector:

- distribution obtained from the first N- dN samples of the matrix vector.
- distribution obtained from the full matrix vector (N samples).

This is done by a simple classification matrix. The test then the test is chi square with the following entry:

- specified threshold stability (between 0 and 1);
- reference distribution: distribution obtained from the Matriz N;
- proven distribution: distribution obtained from the matrix N-dN.

Depending on the result of the Kolmogorov-Smirnov test, if the value is greater than the stability threshold, it is considered that the assessment of stability has been successful.

Calculation of device interference

The ICE has two distinct functions:

- Process different interfering signals to calculate the probability of interference. Three types of interfering signals are considered: emissions essential, out of band emissions, blocking and intermodulation.
- Get generic limits. The output of the ICE is, therefore, one surface multidimensional where the probability of interference based on radio parameters shown. The general organization of the ICE shown in Fig. 19.

Distributions are calculated interfering signals with respect to reference levels or functions of unwanted attenuation (template emission), lock (template reception) or intermodulation. Law translation of the reference $cdf^{ref_{i-init}}$ reference to ^{ref_i} It is given by the following formula:

$$P(iRSS_i(ref_i) < X) = P(iRSS_i(ref_{i-init}) < X - t(ref_i - ref_{i-init})); \quad t = \begin{cases} 1; & i = spur \\ -1; & i = block \\ -3; & i = intermod \end{cases} \tag{8}$$

The complete and rapid flowcharts ICE (approximately) shown in FIG. For simplicity, in the flowcharts of FIGS. 20 and 21 shows the case of t • 1 (equation (8) spur case (not essential)).

Quick calculation algorithm

In the algorithm ICE quick calculation of the following two assumptions are made:

- The iiRSS are independent variables in which the index i corresponds to the ith type of interference situation.
- One of the iiRSS It is dominant over all other interfering signals.

Undesired overall probability PD composed of not being interfered by the signal is:

$$PD = P(\frac{dRSS}{iRSS_{composite}} > \frac{C}{I} | dRSS > sens_{vr}) \tag{9}$$

Using the second hypothesis, we may approximate equation (9) by the following equation:

$$PD = P(\prod_{i=1}^{n} (\frac{dRSS}{iRSS_{composite}} > \frac{C}{I} | dRSS > sens_{vr})) \tag{10}$$

and as iiRSS are independent variables, we can write the equation (10) as:

$$PD \sim \prod_{i=1}^{n} P(\frac{dRSS}{iRSS_{composite}} > \frac{C}{I} | dRSS > sens_{vr}) = \prod_{i=1}^{n} P_i\left(\frac{C}{I}\right) \tag{11}$$

Each interference situation has a set of references, refi, eg spur, avr, etc. The user can choose the set of references to be used in calculating PD. Introducing refi in equation (11), the following approximation is obtained:

$$PD = \prod_{i=1}^{n} P_i\left(\frac{C}{I}\right) \qquad (12)$$

which it is used in the quick calculation algorithm. One can easily show that 1 - PD is the probability of occurrence disturbance caused by at least one of the n interferers.

ICE entire chart

Three cases are considered:

The desired and / or interfering signals are correlated. In this case, the probability PD is calculated by processing the data vectors directly. For each situation of interference, interference signals for all interfering sources iRSScomposite add. Then, from the two vectors and iRSScomposite IrSS probability PD is determined:

$$PD = P\left(\frac{dRSS}{iRSS_{composite}} > \frac{C}{I} \middle| dRSS > sens\right) \qquad (13)$$

adding all the terms that meet dRSS> sens. Similarly to the case, the fast calculation, when elements of the data vectors are added to calculate (12) mode, data should be updated so that they correspond to a desired set of references.

All signals are uncorrelated and their distributions (calculated by the DEE) are given by a closed figure. First, the build-up and distribution functionative interferer composite signal is calculated by integrating the functions of iiRSS distribution. Note that the refi cause linear displacement of the iiRSS distributions with respect to each other. In calculating composite iiRSScomposite must move the iiRSS distributions so that all are related to the same set of references. Finally, it is calculated (12) using the formula of conditional probability that integrates dRSS and iRSScomposite distributions.

The third case It is similar to the second but with the difference that iRSScomposite distribution function is determined by the Monte Carlo technique.

The flowchart of FIG. 19 describes the logical process of the ICE, which is suitable in the case of full integration for calculating iRSScomposite (see flowchart of FIG. 20). However, in the case of a vector of input data or a process of Monte Carlo sampling, calculating the sum of the vectors to determine iRSScomposite and testing iiRSS respectively, consuming time and resources significantly only can performed once.

Note 1- The calculation time is the criterion to choose between sampling or integration.

Note 2- This formula is detailed in Document SE21 (96) / 20. (DRSS / I) Is the criterion used in this example. Other criteria may be used.

Note 3- ref1, ..., Refn are the values of the relevant parameters (spur, AAV, ...) which is necessary to calculate the probability of interference.

Note 4 - The meaning of this sum is symbolic since it must make the sum of the linear and iiRSS values expressed in dB.

Organigrama de cálculo rápido del ICE

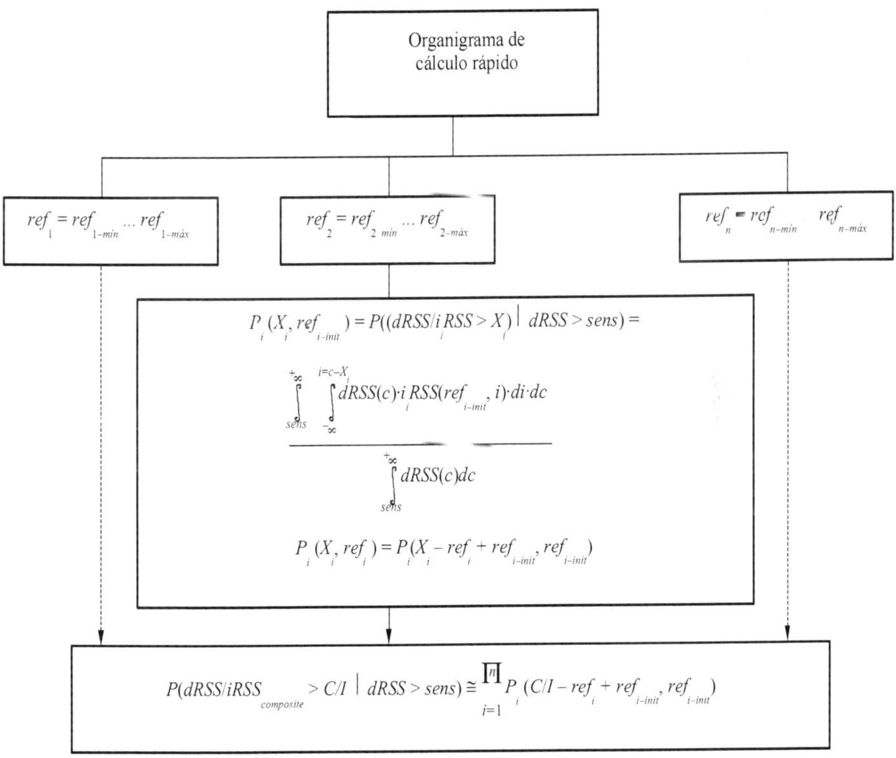

2 ALTERNATIVE MODELING OF APPLIED MATHEMATICS CONTEXT IN PHYSICS EDUCATION: INTERDISCIPLINARY RELEVANCE BETWEEN MATHEMATICS AND PHYSICS

Ivan Stanimirović

INTRODUCTION

The practical problem solving in mathematics in the school context often restricted to "problems - type" whose goal is to equip the resolution mechanisms through repetition of these problems.

This situation is no different in Physics Teaching in High School. In this case, the resolution of problems is mostly the application of formulas with no apparent relationship to the physical concept, consisting in machining procedures, spreading what has been called "mathematization" of the Physics Teaching. This process of "mathematization" checked in our schools in Physics classes, characterized by excessive emphasis on ownership of mathematical concepts to solve physics problems, unrelated to the physical phenomena under study.

Paty (2003: 1) studied the mathematization process of the physical world operated in the eighteenth century, very different from this that we see in our schools, highlighting the "D`Alembert role as a central character in this process, as it has refined and reshaped concepts, both epistemological as methodological important for the mathematization of the physical world. " Although the textbooks bring contextualized problems related to the daily life of students, there is a need for teachers to revise their teaching in order to make problem solving a challenging activity that mobilizes the cognitive process of students, leading them to develop their critical capacity, in order to discuss the developed mathematical models, validation, if there are other models for the same phenomenon, among other issues to be raised. Moreover, the symbolic language must be worked from the early grades so that students can understand significantly its relationship with the mathematical models. It must be clear that the mathematical models developed in the Physics Teaching have an added meaning to the phenomenon under study, their variables have a meaning and does not constitute a set of empty symbols.

MATHEMATICAL MODELING AND SYMBOLIC LANGUAGE

To Bassanezi (2002: 16) "mathematical modeling is the art of transforming reality of problems in mathematical problems and solve them playing their solutions in real-world language."

This "art of transforming reality of problems in mathematical problems" requires a language which is symbolic language.

NCPs (1999: 237) of Mathematics and Physics emphasize the use of symbolic language, as you can see: "(...) to express themselves correctly using the appropriate physical language the elements of its symbolic representation and clearly display and objective knowledge learned through such language. "

However, since the initial series has not had an effective working with the symbolic language. In general, the first contact of students with such language is in the 5th grade students to the study of set theory, when this content is inserted into the teaching plan. The second contact is in 6th grade when students learn equations. In this case, the student is introduced to the world of mathematical modeling by solving problems that require the assembly of equations and, in general are in problems-type.

In 7th grade, with the study of monomials and polynomials, the student has an effective contact with the symbolic language. There is a great opportunity for the teacher to work algebraic mathematical modeling through problems involving geometry concepts, a focus intradisciplinary.

Let us be attentive, to the fact that it takes work to symbolic language from the initial series of basic education when the student starts the development of mathematical knowledge when assimilates the fundamental operations and can solve problems of addition and subtraction, developing their capacity for abstraction and representation.

Thus, it is clear that the symbolic language is a key element in the mathematical modeling. Bassanezi (2002: 17) asserts the relevance of symbolic language for Science:

"Science as accumulated knowledge, depends on coding and symbols associated with oral or visual representations communications (common action to understand, explain and manage the reality) giving rise to language and representation." He continues, emphasizing that the mathematical model "is a set of symbols and mathematical relationships representing somehow the studied object. "

However, at the beginning students will face difficulties with the use of symbolic language, but it is essential to instill the idea that it takes patience to build the rigor of mathematical language. When we apply such language to mathematical models in physics education, this takes on a sense related to the phenomenon under study, no longer a static set and now represents a web of meanings. So there is a direct relation between mathematics and physics as he lay in Bassanezi (2002: 19-22):

"In scientific research, mathematics began to function as a unifying agent of a rationalized world, being an indispensable tool for the formulation of fundamental phenomenological theories, mainly due to its power of synthesis andgeneralization. "It further states that" The fundamental laws of physics are formulated mathematically to provide a first generation of mathematical models that later are subject to a number of corrections, some empirical. "

This relationship between mathematics and physics becomes as essential then emerging physics - mathematics, constituting as stated

Paty (1995: 234-235) "a physical measurement itself, that is, the explicit elaboration of thought physical concepts mathematically: mathematisation conceived as being inherent to the concepts, these constituent, used to construct them. (...) Since then, physics has replaced the real determinations given in experience for these abstract concepts built. " Hence denoted as asserts Lozada et al (2006: 3) that physics consists of a triad ". Theory formulation, preparation of a consistent mathematical model and experimentation" is perceived then be the experiment a factor that provides consistency to the mathematical model of the physical concept.

According Paty (1995: 237) mathematics is to physics "a means of exceptional and constituent research, for without it there would be, in particular, this unitary vocation that is as physics, the material universe science."

Notes on problem solving in Physics

Troubleshooting in Physics has been much criticized, especially by reducing to the application data in formulas or algorithms, not establishing any significance to the physical concept. Moreover, it created a "culture of problems - type", which are not challenging and create the "mechanization" of problem solving.

In this sense, Peduzzi (1997: 230-231) cited Zylbersztajn (1998: 2) states that: "Particularly in physics teaching area (...) what is happening is that the teacher, to illustrate the problem solving, promotes a linear resolution, explaining the situation in question as something whose solution is known and does not generate any questions or require attempts. That is, he treats problems as illustrative, as a theory of exercises rather than real problems, which is what they represent for the student. "

This view is supported by the PCNs (1999: 229) that indicate that the Physical Education "emphasizes the use of formulas in artificial situations, separating the mathematical language that these formulas represent their actual physical meaning. Insists on solving repetitive exercises, claiming that learning occurs by automation or memorization and not the construction of knowledge through skills acquired. "

The Curriculum Guidelines for Secondary Education (2006: 54) confirms what was said by the NCPs: "(...) In practice it is common to problem solving using mathematical expressions of physical principles and the model used. This is due in part to the fact already mentioned that these problems are so devised so that can be solved with the mere application of formulas, just the student know what expression to use and replace the data in the statement of the problem. "And continue, warning that "these practices do not ensure the investigative competence, because it does not promote reflection and knowledge construction. That is, in this way it is taught badly and learn it worse. "

In our schools, this practice is frequent. Students often say that "in Physics classes studying mathematics," because there is no relationship between the physical concept and the mathematical model, by the way there is the practice of solved problems in physics with the goal of developing a mathematical model. There is an illusion that students "solve problems", since mecanizam resolution procedures. The results are even questioned and analyzed in order to discuss the phenomenon studied. Solving the problem ends when you get a numeric result. In this case, is there a significant learning?

Larkin, in turn, conducted studies comparing the way beginners and experts solve problems and found that beginners usually resort to the formulas and those skilled usually from a physical representation, followed by preparation of a scheme and in accordance with this solves the problem, working from the statement to the unknown.

Kuhn (1975) defends the "problems - type", stating that these are fundamental in learning physics. Zylbersztajn (1998: 7) explains that in Kuhn's vision students to solve exemplary problems learn to apply the appropriate versions of the laws (symbolic generalizations) to specific contexts, a process through which new problems come to be regarded as similar cases to those previously found. Kuhn (1975: 235) cited Zylbersztajn (1998: 8) emphasizes that this process of similarities also occurs in the history of science: "Scientists solve puzzles by modeling them according to previous solutions often with minimal resources, the symbolic generalizations. "

These "problems - type" are criticized by Gil - Perez et al (1998, 1992), because according to the author, lead students to adopt a "mechanical operativismo".

A survey conducted by Sousa and Favero (2003: 19- 20) pointed out that "at least half of the subjects see the problem solving as application of the theory, ie, application and acquired knowledge transfer." And continuing: "The student does not become aware that the relationship between theory and practice is interactive, so it is normal to have difficulty solving problems, as this is also responsible for learning theory. Faced with this lack of awareness, the student uses the mechanical use of formulas as if there was the application of the theory that it can not assimilate. "

Considering the Meaningful Learning Theory of David Ausubel, there are many points to be observed in relation to learning through problems - type as the author states that there rote learning and that it is inevitable in the case of entirely new concepts for the student and then it will become significant. Necessary to carry out a study more deepened this aspect in relation to problem solving in physics to see if this further processing is a significant fact, which the intensity and changes occurring in the student cognitive process. So it does not intend to - to exhaust the debate on this issue.

It should be noted, however, that the way to solve problems is being worked on Physics Teaching, it is not providing opportunities for the development of mathematical models, since we know that physics is a science that works with theory formulation, preparation a mathematical model and experimentation.

There is no correlation theory about the physical phenomenon and its mathematical model. Meanings are not assigned to the "formulations". Thus, the Physics Teaching becomes empty, meaningless, undermining the acquisition of knowledge, which according Skovsmose (1990) cited by Barbosa (2001) is set in three different types of knowledge that can be related to mathematical modeling:

- Mathematical knowledge itself;
- Technological knowledge, which refers to how to build and use a mathematical model;
- Reflective knowledge, which refers to the nature of the models and the criteria used in its construction, implementation and evaluation.

The significant knowledge is one endowed with meaning that makes sense to the student and mobilizes their cognitive process.

THE THEORY OF FIELDS OF VERGNAUD CONCEPT: THEORETICAL CONTRIBUTIONS TO MATHEMATICAL MODELING APPLIED TO PHYSICS TEACHING

The Theory of Vergnaud's Conceptual Fields as well as being related to problem solving, is also directly linked to mathematical modeling as will be shown.

According to Vergnaud (1990: 133) CBT is a cognitive theory which aims to provide a coherent framework and some basic principles for the study of development and learning of complex skills, especially those related to science and techniques. For Vergnaud (1983: 40) knowledge is organized into conceptual fields "are an informal and heterogeneous set of problems, situations, concepts, relationships, structures, content and thought operations, connected with each other and probably intertwined in the process of acquisition "to this theory, the study of the specificity of the content involved in the teaching -. learning is essential, because through them can analyze the difficulties presented by the students. Another important aspect pointed out by Vergnaud concerns the conceptualization and in this sense to develop mathematical models applied to Physics Teaching requires physical concepts are seized so that relations are established and work out the mathematical model. Situations, according to the author, give meaning to the concepts and to this end, mathematical modeling is inserted in this case, since it allows to model phenomena related to the daily life of the student.

Ferruzzi et al (2004, 1357) exemplifies the mathematical modeling from the operation of an electric oven. Using this example and by study of the triple Vergnaud, related to the mathematical modeling applied to the Physics Education:

S: is the set of situations that give meaning to the concept. For mathematical modeling applied to physics education, one can quote a problem situation, which in this case is to understand the functioning of an electric oven;

I: I set of invariants on which rests the operation of the schemes (meaning). For mathematical modeling applied to Physics Teaching in the case in question, "to the proper functioning of the oven, temperature control is necessary, that is, it is necessary that the oven is turned off when it reaches the ideal temperature for a given product and is switched off again if there is a big temperature drop. "

A: all the forms of language and not language that allow symbolically represent the concept, its properties, the conditions and treatment procedures (signifier). For mathematical modeling applied to the Physics Teaching in said example is presented for the latter triple element of the mathematical model representing the behavior of the temperature versus time: T * - Tn = a. and bn

Thus, as emphasized by Ricardo et al (2003: 6) "The development and operation of a concept always involve the overlapping of these three dimensions" and that could be understood by the example cited above, which begins with the initiation a contextualized problem setting off on a mathematical model, showing the direct relationship problem solving with mathematical modeling.

QUALITATIVE RESEARCH AND SUGGESTIONS FOR THE DEVELOPMENT OF MATHEMATICAL MODELING APPLIED TO PHYSICS TEACHING IN HIGH SCHOOL

In June 2006, in a Physics Teacher Training Workshop (Lozada et al, 2006), a qualitative research was carried out in survey mode (Fiorentini & Lorenzato, 2006: 106-107) in order to establish some aspects relating to the training of teachers of physics and their teaching practice (methodologies, content and work with students to develop as well as the difficulties that students have in the process - learning in Physics).

The research was conducted through a questionnaire divided into four categories of analysis, namely: Profile Teaching, Teaching, Teaching of Physics and Particle Physics. Of the 70 enrolled, 40 were teachers of which 23 attended the workshop and 21 completed the questionnaire.

Within the four categories of analysis, there is the category "Teaching" in which among the questions asked two questions deserve attention, with respect to assimilate the contents and the other in relation to the difficulties presented by the students. The data analysis is extracted that between difficulties pointed out, we stress the first three that are directly related to the development of mathematical models in physics because they require cognitive mobilization (assimilation of the content and interpretation of statements / texts) and mathematics tool (mathematical operations that involve the establishment of models mathematical).

By analyzing the results, it was found that 80.95% of teachers said their students have the contents assimilation difficulties; 95.24% of teachers said their students have difficulties in the interpretation of statements / texts; 80.95% of teachers said their students have difficulties in mathematical operations. Regarding the writing, 38.10% of the teachers pointed out that this is also a factor that hinders the learning and 9.52% said that their students have other difficulties such as logic / reasoning and pointed disinterest as a factor that hinders the Learn. Conceptually texts of Physics that the answers to questions do not require operations, writing is a key factor, because through it the students express their ideas,

Preliminary considerations

While considering that the sample studied cannot be understood as full representative of the universe of teachers, through a preliminary analysis the information collected it can be seen the difficulties of students identified by teachers in relation to the mathematical content used in physics and consequently interfere in the resolution of problems and the development of mathematical models.

Thus, despite the teacher sample was very restricted, the data clearly show the concern of the same in relation to the teaching and learning processes in physics, fitting to point out that of course there is the need for a deeper study to check out other causes of low student performance in mathematical modeling in physics.

Because the search results, he continued to research, seeking alternative solutions to mathematical modeling applied to Physics Teaching. Thus, it is considered important and necessary to carry out interdisciplinary work between mathematics and physics. The integration between the two disciplines brings several pedagogical advantages, including the ability to relate content. Also, for the mathematical modeling applied to physics education in high school it is suggested:

a) Mathematical modeling by bringing problems: it is proposed problems to a certain content of Physics and that are contextualized, linked to the daily life of the student. In this regard, the Curriculum Guidelines for Secondary Education (2006: 60-61) are positioned, "The first step in a contextualized learning may come from the choice of phenomena, objects and things of experiential universe. Real-world problems tend to provide often more creative solutions and are presumably more meaningful and motivating to artificial problems. It is interesting for the students can bring the abstract world of physics to the world built daily in their experiences. "Thus, discussion of the problems for the development of the mathematical model allows to develop critical, the reasoning capacity, the selection-solving strategies, and allows develop a collective construction of knowledge work, developing values and attitudes. Moreover, you can choose to contextualized problems under the CTS approach (Science, technology and society), allowing you to develop critical attitudes toward social events involving scientific and technological knowledge (2006: 63). It must be said, it is extremely important to work with non-numerical problems, those requiring only conceptual answers. In this case the theory of mental models (Laird, 1993) proves to be suitable for understanding how students construct conceptual models. developing values and attitudes. Moreover, you can choose to contextualized problems under the CTS approach (Science, technology and society), allowing you to develop critical attitudes toward social events involving scientific and technological knowledge (2006: 63). It must be said, it is extremely important to work with non-numerical problems, those requiring only conceptual answers. In this case the theory of mental models (Laird, 1993) proves to be suitable for understanding how students construct conceptual models. developing values and attitudes.

b) Moreover, you can choose to contextualized problems under the CTS approach (Science, technology and society), allowing you to develop critical attitudes toward social events involving scientific and technological knowledge (2006: 63). It must be said, it is extremely important to work with non-numerical problems, those requiring only conceptual answers. In this case the theory of mental models (Laird, 1993) proves to be suitable for understanding how students construct conceptual models. which is of fundamental importance to work with non-numerical problems, those requiring only conceptual answers. In this case the theory of mental models (Laird, 1993) proves to be suitable for understanding how students construct conceptual models. which is of fundamental importance to work with non-numerical problems, those requiring only conceptual answers. In this case the theory of mental models (Laird, 1993) proves to be suitable for understanding how students construct conceptual models.

c) Mathematical modeling by computing: adoption of software such as Modellus (Araújo, Veit and Moreira, 2003; Reis, 2005; Veit, Mors and Teodoro, 2002) and Stella (Santos, 1998), highlighting the importance of pencil and paper modeling and considering these programs as an important tool in the modeling process.

d) Mathematical modeling through experimentation: the PCNs (1999: 12) state that "the student must interpret and criticize from experiments and demonstrations", emphasizing experimentation as a resource that can be used for physics education. In this case it is necessary for the teacher to draw up a questionnaire survey of students' prior knowledge related to the subject of experimentation. Another questionnaire must be answered after the experiment so that we can verify the assimilation of content, and it is suggested that students prepare a report describing the experiment conducted, this timely moment to work writing.

Thus, the experiment is presented as an example, operation of an electrical circuit (Ferruzi et al, 2004: 1356). This experiment will provide:

- Checking whether there is any relationship between the voltage, electric current and resistance of a material;

- Determine a mathematical model which describes the behavior of the current flowing in a circuit in relation to the applied voltage and the resistor equipment;
- Entering the concept of function of the 1st degree, defining linear and angular coefficient, which shows the feasibility of interdisciplinary work with Mathematics and the multidisciplinary nature of mathematical modeling (Bassanezi, 2002: 16).

However, it must be clear that you do not want to perform the experiment spread the empiricist view of science. The goal is to bring the student to develop a mathematical model existing and consolidated by a theory. Silveira and Peduzzi (2006: 50-52) state:

"There is no doubt that experiments, observations, outcome measures are important for scientific knowledge. (...) The production of scientific knowledge can not be understood through the empiricist epistemology (although the books - texts and many scientists believe so) and can not be described as a result of applying a scientific method that begins with observational / experimental results . "and continuing:" the story (or caricature) empiricist not only impoverishes the history of science, leads to distorted views of the nature of science and the scientific enterprise. "

Modeling through experimentation is a way to work with the open laboratory. According Capechi (cited Carmo and Carvalho, 2006: 4), the open lab consists They claim that this time is ripe for teacher and student articulate the language "an experimental activity that part of the problem raised by the teacher and involving students." mathematics used by physicists with other languages to build the scientific meanings. This is the perspective of education for research (Capechi, op. Cit), which supports the principles of meaningful learning and contributes to activate the basic ideas that the students have. In this step, to develop mathematical modeling through experimentation, it is suggested that students develop conceptual maps (Novak, 1998) establishing the relationship between the various concepts that involve the phenomenon.

Moreover, the phenomenon needs to transpire in the mathematical model needs to be "seen" in the representation of mathematical language. There must be a link between the typological features and the topological features such as points and Carmo Carvalho (op. Cit).

Bassanezi (2002: 26) cites the trial as one of the intellectual activities of mathematical modeling. For the author, this intellectual activity also includes the following steps: abstraction, resolution, validation and modification of the mathematical model.

However, it is not always possible to perform mathematical modeling of physics content in high school, such as some contents of Modern Physics and Contemporary, as is the case of Physics of Elementary Particles (Lozada & Araújo, 2007). Here is an epistemological obstacle to the didactic transposition of this content to the high school, which is usually addressed in the conceptual field of historical-chronological order. One suggestion to overcome this obstacle would be to select topics of Modern and Contemporary Physics related to elementary particles such as atomic models. Thus, the mathematical modeling of this content could be implemented by the Bohr formula for the hydrogen atom (= En - 13.6 / N^2). (Ramalho Jr et al, 2005: 406-408).

FINAL REMARKS

Although the context in which it occurs is another physics, mathematics is a constant support tool, particularly in the resolution that culminate in mathematical models.

In this regard, Barbosa (2003: 5) states: "Despite the situations have originated in fields other than math, students are invited to use any ideas, concepts, mathematical algorithms to address them."

In addition, urgent to do a job with the symbolic language from the early grades, so that students can develop their capacity for abstraction, ending up in a study of the linguistic field adopted in mathematical modeling. Moreover, it is necessary to follow the routes that students follow when developing modeling, checking how students develop modeling, his advances and setbacks, in order to improve the modeling process.

Problem solving is suggested that oportunize to the student contact with everyday situations so you can develop mathematical models, including its added meaning to the physical concept. The teacher should establish an interdisciplinary work with mathematics in order to minimize the difficulties with the mathematical tools used in mathematical modeling applied to Physics Teaching and implement teaching strategies such which that have been suggested in order to develop the skills and competencies students in modeling, providing a challenging education that stimulates the cognitive development of students, leading them to a more effective participation in the learning process.

Not withstanding the discussions on the resolution of problems and their relation to mathematical modeling, there is a consensus regarding the mathematical modeling be considered learning environment for the teaching of Mathematics (Bassanezi, 1994; Blum, 1995; Borba, Meneghetti and Hermini, 1997). This idea is supported by Barbosa (2001: 5): ". Modeling, to me, is a learning environment in which students are invited to discuss and investigate, by means of mathematics, situations with reference to the reality"

And this reference in fact in daily life is marked by Burak (1992: 62):

"Mathematical modeling is in a set of procedures whose goal is to build a parallel to try to explain, mathematically, the phenomena present in everyday human being, helping you to make predictions and decisions."

Thus, the daily enables providing different situations for a dynamic work in mathematical modeling in the room, especially because students can observe the world around them and to model physical phenomena, involving an interdisciplinary approach between physics and mathematics, as previously asserted. The considerations set out here do not exhaust the debate on mathematical modeling applied to physics education, and research continue, aiming to identify other aspects and referrals to the realization of mathematical modeling applied to Physics Teaching in High School.

3 SURVEY OF INTERDISCIPLINARY TOPICS IN COMPUTATIONAL SCIENCE

Ivan Stanimirović

INTRODUCTION

Basic training is meant to introduce the materials necessary for the technological development of computing. The main ingredient of this area is the computer science featuring egress as belonging to the area of computing. Most materials are technological applications of computer science. Are matters of basic training of computing area courses: computer science, mathematics, physics and electricity and pedagogy.

Computer Science

The central point of this material is in the machine concepts and algorithm. According to the authors classics of computer science, algorithm is a set of instructions in a language, played by a real or abstract machine. Given a machine and a problem, the solution is given by an algorithm. You can not then separate the concept of machine concept of the algorithm. No machine there is no algorithm. An egress a course of reasoning computation differently from other professionals because it has the ability to build algorithms such as troubleshooting. The Computer Science is the most important area in the composition of the curricula of the courses and therefore is directly related to the objectives of the training of human resources. The sub-areas are the following.

Programming

The program, understood as computer programming, is an activity aimed at solving problems. In this sense it is related to a wide range of other activities such as specification, design, validation, modeling and structuring of programs and data, using programming languages themselves as tools.

Contrary to what was proclaimed a few years ago, the programming activity has ceased to be an "art" to become a science, involving a set of principles, techniques and formalities aimed at well structured and reliable software production. Look up, among them, the principles of abstraction, encapsulation and modularization techniques and structured programming.

Therefore the study program is not restricted to the study of programming languages. Programming languages are in a software implementation tool, which is the result of applying a series of knowledge that transform the specification of the solution of a problem on a computer program that effectively solves that problem.

In the study of programming languages should be given emphasis on the functional and structural aspects of programming languages over the syntax details. Concepts such as the meaning of association, evaluation, assignment, procedure call, messaging, parameter passing, inheritance, polymorphism, encapsulation, etc. They should be emphasized. The study of languages should be preceded by the study of the major programming paradigms, notably the imperative programming, functional, based on the logic and object-oriented.

The development of algorithms, together with the study of data structures should receive special attention in the theme programming approach. It also should be given emphasis to the study of specification techniques, design and validation programs. Excellent course for the exercise of programming consists of the research study in tables and sorting techniques.

Computation and Algorithms

Computer programs (or "software") are founded on three basic theoretical concepts: algorithms, computer models and formal languages. An algorithm is an abstract method but well defined for solving a problem in finite time. The notion of algorithm assumes the existence of some kind of abstract machine where it can be run automatically. We call "computer models" different abstract machines on which the algorithms are formulated. The bridge between these two concepts is the concept of formal language, which allows the expression of a given algorithm for a particular computing model; this expression is called the "program."

The study of computer algorithms and models allows to address the following fundamental questions: What are the theoretical limits of what can and what can not be solved through computers (ie what is computable)? Within what is computable, what are the most efficient algorithms and data structures? How to characterize the efficiency (or complexity) of the algorithms? How can design and analyze an efficient algorithm? It should be noted that the high abstract level at which these studies are done provides conclusions that go beyond the dizzying technological change they are experiencing modern computers.

The study of syntactic and semantic aspects of formal languages is fundamental to programming activity, since all programming languages are formal languages. In addition, there are in many other situations computing using formal languages. A good example is the concept of regular expression, which appears frequently in word processing.

Computer Architecture

The term computer architecture refers to existing features in a machine design to perform the tasks written in any programming language (the study of machines running programs, or computers). The knowledge of this area is essential not only for those who will design new computers, but also for those who will use them. Knowledge of the basic principles of operation of computers and embedded technology in these allows for more efficient use of resources and the determination of the classes of problems that can be solved with this technology. The design of a computer involves several aspects including:

a. Set of instructions
b. functional organization
c. logical design
d. Implementation

The architectural design aims to optimize a machine along these levels. The instruction set is that which is visible to the programmer (or compiler) in the development of programs. Defines the various forms of data addressing specific capabilities for handling for some data structures and instructions that can compose a program. The set of instructions form the boundary line between the hardware and the software, requiring knowledge of basic software for hardware design. The specificity of a given instruction set can generate optimized machines to process a certain specific type of problem.

The organization provides functional blocks necessary materials to the interpretation and execution of the instruction set. Classically a processor is divided into Control Unit, data flow and memory system. It notes that although this division of functions is widely used, is not the only possible functional partitioning to be used. It includes the high-level aspects of computer design, as the system memory, the structures of buses and communication with peripherals and internal features of the central processing unit. Techniques used as buffering instructions, pipeline and other are included here. In the functional organization are also the working principle of the various peripherals and their communication with the processing unit. (Includes up here handlers breaks

The logical design refers to the design of the various functional elements in digital logic, such as arithmetic operations (Arithmetic Logic Units) and algorithmic systems that are embedded in the processor (as treatment interruptions) and the various component elements of the processor, memory and peripherals. Elements of sets algebra, in particular Boolean algebra and logic design and optimization techniques are included here. automatic synthesis techniques belong to this domain, and an idea of them contributes to the understanding of the project quickly and new implementations that appear on the market. For the control units, the interpretation techniques levels are included herein as conditional controllers, VLIW, and classic microprogramming among others.

The implementation includes integrated circuit designs, in various technologies, power consideration, packaging and prototyping. Implementing interfaces with the area of electrical engineering, generating technologies that enable this implementation.

Optimization of an architecture requires familiarity with specific areas of techniques, such as performance evaluation, operating systems, techniques and systems and digital circuit design.

Mathematics

Mathematics, for the computing area, should be seen as a tool to be used in the formal definition of computational concepts (languages, automata, methods, etc.). Formal models allow you to set their properties and scale their bodies, given their boundary conditions. Whereas most of the computational concepts belong to the discrete domain, discrete mathematics (or also called abstract algebra) is heavily used. Mathematical logic is also a key tool in the definition of computational concepts. Category Theory has buildings whose power of expression does not, in general, parallel to other theories. This expression allows formalize simplest way to more complex ideas and provides a new or better understanding of the issues related to all Computer Science. As Category Theory is a new tool to exemplify, it is worth drawing a parallel with the Pascal language: Category Theory is for sets of theory as well as Pascal is for Assembler languages.

Many computational concepts are based on mathematical models well known as graphs and interval arithmetic. The combinatorics is the basis of optimization algorithms to study combinatorial problems, such as problems on graphs.

Math on the real, continuous mathematics (differential and integral calculations, linear algebra, analytic geometry, numerical calculations, etc.) is important in specific areas of computing. In the areas of operating systems, networks, complexity of algorithms, computer graphics, image processing, simulation, physics, electricity and electronics etc. the mathematics of the continuum is a greater or lesser extent employed. The statistical area has applications in their own area of computing (networks, operating systems, etc.) as the actual problem solving involving computing applications.

Physics and Electricity

Physical, especially electricity concepts, is a tool used in computing, with two main purposes:

- Give the science of mathematical and statistical models used in understanding the phenomena that occur in computers and the interconnection of these.
- Enter the scientific view, where the models try to express the observed reality.

This enables the egress abstract models to work with, fundamental in computing and understand the technological advances achieved through the use / development of new models.

Relevant aspects of the physical area that should be included in the curricula can be classified into the following topics:

a. basic laws of Electricity
b. mathematical representation and Electrical Quantities Measurement Units
c. Operating principle of semiconductor devices
d. Electromagnetic Theory and Wave
e. optical Phenomena

The basic laws of electricity are intended to provide an understanding of the phenomena and problems involved in the technological evolution of the realization of computing machines. The basic laws of current (us) and voltage (mesh) also give the necessary understanding for the limitations of physical connectivity, such as buses and networks between computing subsystems.

The mathematical model of the electrical, furthers our understanding of modulation and interference phenomena involved in various data communication processes, pattern recognition and digital signal processing, these widely used in the fields of hybrid application such as robotics and biomedical.

The electromagnetic field theory and waves and operation of semiconductors enable understanding of the current realization of devices that implement the computational logic as well as the limitations of current technology and the coming years. Besides, when viewed more deeply, allow the design of computing machines (VLSI design and programmable logic), the notion of the phenomena involved in semiconductor technology and waves enables graduates to analyze the technological breakdown processes that will come in the coming years .

In the field of optics, the concepts of reflection, diffraction and attenuation of certain ranges of the light spectrum, enables the future professional to understand the limits involved in optical communications and future in performing the computational logic based on optical principles.

The depth of knowledge presented varies from the core activity of the professional. Those dedicated to the design and implementation of systems should have a wider range of these areas, depending on the specific technological area of expertise (eg microelectronics, automation, data communications). For professionals working in technological areas where the base is the programming or the theory of computation, the understanding of these phenomena gives able to monitor technological developments and glimpse the great moments of paradigm shift in the construction and implementation of computer systems.

Pedagogy

Routinely translated as mastery of techniques, skills and methodologies, aiming at the transmission of a certain knowledge - education, pedagogy has been consolidated in modernity as "science of education" that performs a systematic reflection on the educational practice. It is integrated to all the so-called "Educational Sciences" there with the instrument specific to the pedagogical action. When referred to educational institutions, pedagogy is conceptualized as a practical configuration aimed at the construction and the production of knowledge and learning. In contemporary language is to say that the Education refers to the practice of politics in class, meaning the political expression of the practice, the soil of an action that is intentional and that involves intervention. In this sense, pedagogical science comes to promoting the didactic and pedagogical-professional conditions attentive to historical and socially constructed nature of those skills and knowledge in a world continually changing. As a theoretical and practical tool for action, is basically constituted in the interweaving of two broad areas or fields, namely: a) cultural, scientific and ethical-philosophical, covering knowledge and knowledge able to contribute to the social context of the action pedagogical and their relations with the complex ways in which the learning and social identities are produced; b) didactic-pedagogic, referring to a basic theoretical and practical knowledge and techniques that enable the understanding of the school and its modern setting; education and their pedagogical devices (technologies, methods and strategies of teaching); school knowledge and curriculum organization. It includes also analysis of the professional culture of teaching and educational policies.

TECHNOLOGICAL TRAINING AREA

With the acquired basic knowledge, this area of training aims to show the application of the same technological development. Technological development on the one hand, aims at creating instruments (tools) of interest to society or technologically strengthen the computer systems to allow building before unviable or inefficient tools.

Operating Systems, Computer Networks and Distributed Systems

Operational systems

Operating systems aim to manage the computer operation to offer its users flexibility, efficiency, safety, transparency and resource sharing

In this context, operating systems can be viewed from two perspectives:

a) as a set of programs aimed at hiding the hardware quirks, presenting users with a machine easier to use, more friendly and safer; b) as a set of programs whose main task is to manage the resources available in order to meet the requests as efficiently as possible, ensuring the sharing and resolving potential conflicts.

Operating systems computing resources are basically grouped into four distinct classes: process, memory, storage (files), input and output. Case management involves concepts of communication, synchronization, scheduling, conflict resolution and context switching. Memory management involves concepts of addressing, memory hierarchies and virtual memory. The file management involves concepts of directories, addressing structure and access, security, sharing (competition) and protection. The input and output management involves concepts of interrupts, devices, interfaces and access controllers.

In the evolution of computer systems and therefore the operating system, we have today a strong demand for systems management for not more than one but a computer network. The Distributed Systems study involves, among others, concepts of computer networking, communication protocols, remote procedure call, group communication, distributed file, name resolution and distributed coordination.

Computer network

The Computer Networks constitute a philosophy of use of computers interconnected by communication systems, can now work together, sharing software of hardware resources and allowing the exchange of information among its members.

Computer networks have emerged from the combination of two technologies: communication and information processing. Thus, the area networks are essentially back to the suitability of new communication technologies, which enable the safe and speedy transfer of information and to the challenge of offering new services that address the needs increasingly sophisticated users.

The continuous evolution of communication technology allows you to carry data at high speeds and over long distances enabling integration service networks carrying different media: text, voice and images. Thus, the networks open doors to offering a wide range of services that meet the diverse areas of knowledge, from simple services such as transferring a file or establishing a connection to a remote system, to more elaborate services requiring multimedia features that enable, for example teleconferencing, distance learning, medical care at a distance, etc.

Basic knowledge in Computer Networks area involve the principles of data communications, by submitting your basic concepts, topologies, concepts related to the transmission and coding information (types of transmission, multiplexing and modulation, switching modes, detection techniques errors, etc.), knowledge of how the hardware and software networks are organized into levels, forming the network architectures. Examples of network architectures should highlight the services, functions of each level and the respective communication protocols; the various types of networks (local, metropolitan and wide area), the service network integration and basic aspects of network interconnection.

additional knowledge of the area can offer an overview of network operating systems; the need to manage networks; of possible attacks and methods for the safety nets and knowledge of how to model and evaluate the performance of computer network systems.

Practical classes are also recommended they can, for example, to familiarize the student with the services, aspects of installation, management and safety nets.

Distributed systems

Distributed systems are systems composed of loosely coupled computers, interconnected by network that provide services and allow access and use of data and shared resources.

The main issues to be addressed in the area of distributed systems relate to distributed algorithms, operating systems and kernels, and programming languages environments, reliability (fault tolerance and data security), database, multimedia systems, real-time systems (with applications, for example in industrial automation, robotics, avionics and automotive electronics.).

The heterogeneity of devices, operating systems, languages and protocols, maintaining the integrity of the information and control access to them, the extent of distributed applications on a global scale networks with a very large number of participants, ensuring the requirements of security and meet the time constraints required by many applications are some of the current challenges in the area of distributed systems. The concept of open systems, the existence of standards for these, the use of object orientation, the tools available to the WEB, the mechanisms for the consistency of the systems, even in the presence of faults and real-time scheduling techniques are some the supports available to meet those challenges.

Currently the area of Distributed Systems has been tightly integrated with the area of Distributed Artificial Intelligence (IAD). The major sub-areas of the IAD, multi-agent systems and distributed resolution of problems, have been used as important tools, both theoretical and practical point of view. This integration occurs to the extent that the use of agents, generally based on a social behavior, allows solving problems in a distributed manner.

Compilers

Compilers are translation tools between languages, keeping the original semantics, such as environments for programming languages (compilers, interpreters, debuggers, profilers, etc.) environments for processing natural language (ortho syntax checkers and translators), tools for compatibility between hardware devices (device-drivers, emulators, cross-compilers, etc.), among others.

Compilers The study should address: (i) the structure of a compiler; (Ii) the analysis of source programs, with the study of the most important methods of lexical and syntactic analysis, semantic, organization of symbols and error management table; (Iii) the tools for automatic generation of the components of a compiler; (Iv) abstract machines and optimization of intermediate code; (V) runtime environment; (Vi) synthesis of object programs, including translation schemes run by syntax, generating machine code and code optimization.

It is essential that the end of the compilers course the student is able to justify the choice of tools, environments, and paradigms used languages and their versions in the development of any software project in-the-small. modularity concepts, mantenibilidade, portability and software costs should be analyzed throughout the course.

Teaching compilers should ensure students applying opportunity of the techniques studied in the development of practical projects of realistic size. Compilers is one of the areas of Computer better formalized, which entails implementation of high efficiency and correction tools.

The compilers matter should be preceded by the study of theoretical concepts of languages and automata, operating systems and computer architectures.

The area of compilers has the ultimate goal of bringing the computer own languages of its users, thus facilitating communication between them.

Database

Current technology has facilitated the activity of collecting and storing data indiscriminately, creating the problem to organize them and manage them properly. The area of databases is to propose solutions to this problem. Nowadays, any entity needs of database systems, which serve as the basis for the development of all applications in commercial environments, industrial, administrative and scientific.

Teaching in databases should consider two main factors: the course material itself and the unusual ability to connect with other disciplines. Topics covered should address issues relating to the data itself (organization, modeling, integrity, storage, integration, distribution and packaging) and database management systems - DBMS (architecture, interfaces, interaction languages, query processing, concurrency control, recovery, security, index, buffer management, and file). Additional topics involving new information processing techniques, using artificial intelligence algorithms.

The visa in databases material allows to bridge with matters of Software Engineering, Artificial Intelligence, Compilers, Human Computer Interface, Operating Systems, Distributed Systems, Networks and Programming Languages. Databases can also be used to motivate examples in the areas of further training.

Software Engineering

Software Engineering comprises a set of mathematical disciplines, techniques (computing), social and management that systematize the production, maintenance, evolution and recovery software intensive products. This is within the estimated time and cost, controlled progress and using principles, methods, technologies and processes for continuous improvement. The products developed and maintained according to the principles of software engineering ensure, by construction, satisfactory quality, adequately supporting its members in carrying out their tasks, they operate satisfactory and economically in real environments and can evolve continuously, adapting to a world in constant evolution.

Teaching Software Engineering undergraduate courses may give rise to various disciplines with different emphases. The origin of these subjects may have different motivation as didactic ratings: managerial, technical aspects, theoretical aspects and experimental aspects. The emphasis can be given at different stages of the development process and software maintenance: requirements engineering, analysis, architecture and design, programming, testing, maintenance, quality assurance and management of the software process. It is important to note that these aspects must be integrated into other disciplines such as: databases, human-machine interface, information systems, networks and various laboratories.

In the business plan it is important the various techniques to measure and resource estimates, cost-benefit analyzes, development planning and installation of teams, process management and software product. At the technical level they should be taught the techniques associated with each stage of the software development process, with emphasis on general principles of engineering methods for requirements analysis and software design, characteristics of the different fields of application programming techniques , technical documentation generation, testing techniques, configuration management and software maintenance.

By teaching them must be ensured that the student assimilate the definitions and the basic principles of software engineering through more conceptual or theoretical subjects. One should also ensure that the student acquires experience in the application of these concepts through practice in laboratories and stages. It is strongly recommended that the student is exposed to a variety of operating systems, management of databases, programming languages and paradigms systems, operating platforms, and tools to support the development of software and documentation.

Multimedia Systems, Human Machine Interface and Virtual Reality

Multimedia systems

The training of professionals able to write multimedia action programs and truly adapt to the currently available computing environment requires a minimum set of undergraduate courses - some fundamentals, other applied - that complement each other and that define a certain knowledge domain within the computer science field. The multimedia computing results from a combination of materials that deal with techniques and concepts of the visual and auditory worlds, such as computer graphics, the sonic computing and building multimedia pieces.

Fix the student the fundamentals of this field is a task that requires a solid background in data structures, object-oriented programming, geometry, linear algebra, physics of light, sound physics and their psycho-physical bases of vision and hearing, It is distributed this background in other disciplines who volunteer graduation.

computer graphics should be presented to the student in its canonical form, so that it can cover the geometric transformations, 3D visualization, modeling objects, color systems, lighting, texture, shading and also the fundamentals animation.

sonic computing - seen as auditory counterpart of computer graphics - addresses the nature of the sonic form, the fundamental algorithms for the construction of sonic forms, digital sound processing techniques, languages for audio synthesis and manipulation of sounds and certainly some rudimentary notions of musical hearing systems and languages in general.

In addition, basics of visual programming, publishing, composition, rhetoric, communication and cognition should be considered, as they provide important information to matter.

Finally, the discipline applied to turn to the construction of multimedia pieces - both on local applications, such as online interactive publications - should involve the knowledge presented in the disciplines above the available technology (currently Java, OpenGL, Midi, JavaSound) to lay the foundation of careful preparation and based programs that bring solutions (outputs) at higher levels preceptivos with regard to a common sense logic of visual and auditory perceptions.

The most in demand of multimedia computing applications are part of the online scientific publication, scientific visualization in general, the instructional pieces or tutorials for any area of knowledge, programs for use in surgical medicine, marketing, art, entertainment , and many others.

Man-machine interface

Computer Area professionals produce artifacts that are aimed at specific audiences with varied technical skills and socio-cultural profiles. Such artifacts should be inserted as naturally as possible in the context of work of its members. For this to occur, the specialist in computer must deeply understand the subliminal structure of the work done by "labor experts" (users) and then analyze the possible technology insertion points based on the obtained profiles (user analysis), assess their implications and redesign the current ways of doing work (task analysis). In this sense, there has been increasingly the concern of computer science professionals on how to make the "marriage" tools and computing environments to users, their tasks and their social aspirations. As has occurred since the industrial revolution in other areas such as "industrial engineering", human factors, ergonomics and man-machine interface also comes in our domain knowledge in general with the names of "Human-Computer Interaction "(IHC) or" Human Computer Interface. "

Human-Computer Interaction can be defined as "a discipline related to the design, implementation and evaluation of interactive computing systems for human use, along with the phenomena related to such use." It refers, therefore, not only to the HC interaction interface issues, but also the theories and techniques of interactive systems design. Such theories are based primarily on the study of users of computer technology and how one influences the other, through the understanding of the working context that the person is carrying out through this technology.

Production of a Human-Computer Interface goes through a series of steps ranging from the design phase "conceptual" interface to the steps of testing "usability" made to end users of the system. These steps are employed in many different techniques and tools, borrowed from various disciplines such as Software Engineering, Ergonomics and Cognitive and Perceptual Psychology.

Throughout the process of developing a user interface, the concern with the "usability" of the interactive system under construction should permeate all activities of the process. Who determines whether an interactive system will be successful or not are the users and they prefer, as a rule, easy system to learn and use, even in a reduced functionality, the systems functionality computationally more "powerful", but with a poor interface with which you need to "duel" all the time to produce something useful. To improve the degree of usability evaluation activities by technology experts and user testing during the implementation of the prototypes are absolutely essential in any process of developing user interfaces.

It is important to emphasize again the importance of contributions from other disciplines, since their influences on interactive system design are perceived in terms of the usability of such systems. The design process should therefore be focused on the user, incorporating the cognitive models that support usability elements. analytical or empirical techniques should be used to evaluate whether the system meets user requirements and your task. One should also consider that there are specific groups - such as children, disabled people and new emerging applications - which have particular needs, different from those of traditional users.

By teaching the concepts involved in the development of interfaces you need to make sure that the student understands the size and importance of the problem of designing and building high degree of usability interfaces, be exposed to different specific models of development, learn to use some techniques and high-impact methods on improving the usability applicable for Computer specialists. Practical experience of project that involves the construction projects and / or prototypes as well as its evaluation in accordance with principles well-established interface design is highly recommended.

Virtual reality

Virtual reality can be defined as an advanced technique of building highly interactive three-dimensional interfaces, using unconventional device input and output.

Your application can occur in several areas of knowledge, using or developing the natural abilities of users to perform operations through immersive three-dimensional and multi-sensory access to virtual environments.

This area involves knowledge of: three-dimensional computer graphics fundamentals, computing platforms, high-performance input multisensory and output devices, software and languages for developing virtual reality applications, modeling and three-dimensional animation, real-time simulation, distributed systems, design interfaces, software development, and human factors analysis.

It is interesting to make use of high-performance equipment, special devices and specific software for the development of virtual environments and applications with three-dimensional interfaces. Furthermore, one should explore the vast material development and virtual reality show, available on the Internet.

Artificial Intelligence

Artificial Intelligence (AI) is the area of Computer Science dedicated to the design and implementation of theories and computational models of cognitive function. The Artificial Intelligence aims to make the machine capable of displaying, in the eyes of an outsider, an intelligent behavior in performing tasks and troubleshooting. Therefore, the IA transcends the boundaries of computer science, interacting with areas such as philosophy, linguistics, psychology, biology and logic.

Knowledge Representation, Reasoning Automation, Troubleshooting, Machine Learning, Perception and Natural Language Processing, understood in broad sense, can be considered fundamental areas of Artificial Intelligence.

The Knowledge Representation comes to models for the organization of knowledge and techniques for the representation and manipulation in computer systems. These models can be symbolic in nature (such as logic, semantic networks, frames, etc.) or not symbolic (such as neural networks, genetic algorithms, Bayesian networks, etc.).

The automation of reasoning includes the study of inference methods by which new knowledge can be obtained, for deriving, from the knowledge available. Among them highlight the logical deduction, non-monotonic inference and Bayesian inference.

The Troubleshooting is dedicated to the study and development of algorithms, with the help of heuristic methods, able to solve, for example, problems considered intractable from the standpoint of conventional computing.

The Machine Learning deals with the development of methods of autonomous knowledge acquisition. Learning methods can be classified into inductive (symbolic in nature), probabilistic, genetic and connectionist (the last three of not symbolic nature).

Perception is concerned with the development of systems capable of transforming the environment information data. An example of this are the odors recognition systems, voices, faces, retinas or fingerprints, to detect movements or textures and interpreting manuscripts and recognize signatures.

Finally, the Natural Language Processing is dedicated to the study and development of techniques and interpretation theories and automatic generation of sentences and texts in any natural language (eg., Portuguese, English, etc.).

Some typical application areas of AI are: Expert Systems, Robotics, Speech and Image Recognition Systems, Games, Intelligent Systems Tutorials, Automatic Translators, Data Mining, Information Retrieval, Adaptive Interfaces, etc. In the context of computer science, there has been growing use of techniques of AI in areas such as Database, Software Engineering, Distributed Systems, Computer Networks, Computer Graphics, Computers in Education, etc.

As a basis for the study of the IA are indispensable knowledge of Mathematical Logic, Theoretical Computer Science, Data Structures, Algorithms and Programming Analysis. Knowledge of programming languages developed according to the logical paradigms, functional and object-oriented is particularly relevant for applications in IA area.

COMPUTER GRAPHICS AND IMAGE PROCESSING CG

Computer Graphics brings together a set of techniques that allow imaging from computer models of real objects (or imaginary) or any data collected by equipment in nature. The application of such techniques is several years widespread for several application areas, namely, CAD / CAM / CAE ("computer-aided design / manufacture / engineering" - design / manufacturing / computer-aided engineering), animation and special effects (for advertising and entertainment), graphical presentation of data (economics, management, statistics) and, later, into three-dimensional data visualization produced by simulation or collected by various equipment such as, scanners and weather satellites.

The study of such techniques comprises modeling processes of objects, the representation of data collected in different ways, the generation of images with varying degrees of realism, among others. It is customary to divide the Computer Graphics according to the size of the treated entities. two-dimensional objects, described in a Cartesian plane, for example, are treated and displayed with various processes of those employed in the representation and visualization of three-dimensional objects. Since data collected or generated from simulations, for example, lead to the use of other process visualization. Equally important for Computer Graphics are aspects of man-machine interaction, since the modeling techniques are fundamentally interactive, which provides a strong interaction with the area of Multimedia Systems,

In general, Computer Graphic study requires the use of concepts of disciplines of mathematics, especially algebra linear, analytic geometry, differential and integral calculus and elements of physics as regards the lighting and motion models.

Image processing

The Image Processing sub-area along with computer graphics, addresses the treatment of pictorial information. Among its main objectives include the development of techniques, methodologies and implementations aimed at representation, image processing and communication.

The image representation of the study comprises the various processes involved in the acquisition, scanning, visualization and mathematical characterization of images through transformations or templates, seeking its efficient processing at a later stage.

On the other hand, the image processing itself covers various topics such as highlighting, filtering, restoration, analysis, reconstruction from projections, compression and image communication.

Due to the emerging aspect of this sub-area, the development of projects and case studies in image processing systems, focused on specific problems in engineering, medicine, telecommunications, etc., are important for the formation of the student. Generally, the issues addressed have a multidisciplinary character, and can use specific concepts from other disciplines such as physical optics, information theory, stochastic processes, artificial intelligence, visual perception, among others.

Practice of computer education

This matter is to apply the basics of pedagogy in computer education for basic and vocational education. It answers the following question: How to teach computing in primary and vocational education. Not yet know the correct way to introduce computer skills. Students learn to count using the fingers. Teaching computing begins with an abstract computing model or a more real model? Methods and computer teaching techniques, whether for adolescents professional purposes in technical courses, whether for general purpose of preparing for work in the 5 to 8 of the 2nd degree series, may not be the same ones used for teaching adult and higher education, even as the laboratories required for teaching must have its own characteristics.

The body of knowledge to be introduced should be flexible. The computer education should consider the existence of machines and algorithms. Hold a "theater" representing machines and showing the various parties working with the help of students can be a simple and didactic way of presenting the operation of a computer. The arithmetic unit, represented by a calculator, memory represented by bins, etc., and using simple language, can work in the theater of the students. One seeks an instruction from memory, interprets and passes to the next that executes the instruction. A "stage" thus assembled shows how a task placed in memory can be performed. It can thus be introduced to the concept machine and algorithm.

Then we can propose issues to reach students who will find one or more solutions (algorithm) that works in the "theater" represented by the students, using the plain language of the machine. At another point, more detailed computer simulation can be used and the process of problem solving in these simulators can be repeated. And in a third stage can be introduced a real programming language and basic notions of software and applications.

Computing professionals should produce tools to meet the needs of society. It is almost impossible to enumerate the facilities introduced by information technology in human activity. Some activities are more frequent, for example, in administrative activities, other more relevant, for example, in a patient monitoring system. So that professionals can interact with professionals from other areas in search of complex computing solutions to their problems, the computer professional should know generally and comprehensively these areas. Thus, the courses should choose a training area of complementary human resources, or a matter of that, and define, together with the corresponding departments, a well-formed cast disciplines and offer their students.

FORMULATION OF HUMANISTIC COMPUTER SCIENCE HISTORY

Knowledge of the historical evolution of the computing area shows how we arrived to the present and allows the egress to know yourself as an evolution of its predecessors.

Training of entrepreneurs is a process of providing professional technical and administrative areas with the concepts and skills to recognize and take advantage of business opportunities, creating and managing successful enterprises, either through the establishment of an enterprise or entrepreneurial activities in departments or centers cost / revenue. This process includes training in recognizing opportunities, resource management, and risk analysis gerenciamente, opening and management of business, business planning, capital leverage, marketing, cash flow techniques and knowledge on standards and legislation for the establishment of an enterprise. Also will be developed skills such as creativity, leadership, teamwork, communication facility, etc.

Entrepreneurship is a new way to make more aggressive, competitive and creative production sector. Your practice can be interpreted as a new industrial policy strategy with a view to development of the country, different, for example, the reserve mercada for the computer.

Computers are so prevalent in our society that its importance is unquestionable. They are changing the way we study, work, play and communicate with each other. The study of ethics in computing is the study of ethical issues that appear as a result of the development and use of computers and computing technologies. It involves identifying and disseminating issues and problems that are within its scope, increasing awareness of the ethical dimension of a particular situation. It also involves studying how to address these issues and problems in order to advance our knowledge and understanding of these problems and suggest solutions to them wise [Johnson & Nissembaum 1995].

The didactic approach to this matter can be quite varied: reading articles, books and articles published in magazines and non-technical papers, discussion of real or fictitious cases, group work on specific topics, interviews with recognized professional competence and reputation, study the codes of ethics of class societies, etc. This matter should be related to subjects such as information systems, computers and society, methods for software development, etc.

Topics covered should evolve as technology evolves and affects the behavior of society. Current topics that may be mentioned are: unauthorized access to computer resources (hackers, viruses, etc.); software property rights (piracy, the current law regulating ownership of software, reverse engineering); confidentiality and privacy; safety; risks of computing and critical systems for security; professional liability and professional regulation; software that discriminates minorities, concerns in the areas of health and environment.

Computer and Society

No machine ceases to have any negative side effects. This matter should be given emphasis to social and individual negative influences caused by computers (the benefits are already widely disseminated). Being abstract machines, and algorithmic, the main effect on its users is to force an abstract thought, logical-symbolic and algorithmic. Secondly, being a machine that simulates human thoughts, and therefore virtual, it does not produce visible disasters, as do the specific machines. One of these disasters is the induction of mental discipline, typical of programmers (basic source of the "bug" of the year 2000 - if the programs had been well documented, would be easily changeable) but also of users who employ eg text editors. In this case, any correction can be made,

A key aspect that should be discussed with students is the influence of the computer on the mentality of developers and users. By presenting a logical-symbolic space deterministic, the computer tends to produce hard thoughts, to be always based on rigorous logic.

From the social point of view you should approach your computer problem replace human labor, especially that dignifies man, and not only the one that degrades (although it is perhaps important to work even if it is not dignified in Instead of creating unemployment by indiscriminate automation). Diginificante a working replacement example is the use of computers in education if this decrease the presence of the teacher.

It is important to make a discussion about the negative effects of the Internet, how to induce the exchange of telegraphic correspondence, the opportunity to publish something without someone to take responsibility for checking the quality, the exponential increase in existing it garbage, the fact that children may have access to information decontextualized, the negative social effects such as isolation, etc.

Finally, ways to counteract the pernicious influences of computers on the minds of its users and developers must be addressed. The practice of artistic activities is an example of possible antidote to compensate for rigid thinking imposed by the computer. In this, creativity tends to be a mere combination of instructions and pre-existing commands and mathematically well defined. On the contrary, in artistic activity the mental space, wholesomely accompanied by emotional, is open and ill-defined.

Sociology

Humanistic instrumentalization and ethics top of the computing and information technology field resumes finds its main justification on the importance for present and future generations, studies, sufficiently contrasted, of modern and contemporary societies, aiming to understand the aspects of social and cultural life to which they belong, in terms of political, cultural, scientific, technological and its values; as well as critical analysis of social relations and its close connections with the revitalization of civic life. Fundamentally the study of these relationships will take generations of professionals to understand the social dynamics and their inclusion in the same, the political interests, structures and power relations in society.

Faced with the challenges posed by technological innovations and changes in work organization is required of the professional of the third millennium the knowledge of trends and organizational conceptions of work, changes in the work content and new requirements for qualifications imposed by new technologies. Such changes indicate the basic principles that should form a proposal for professional training that takes into account the challenges of new technologies and the needs of the population. The specificity of the sociological approach enables the formation of the subject in a polytechnic perspective, which represents the synthesis between general education, vocational training and political education,

The sociological approach can not dispense with the analysis of the new skills needed by professionals in the face of changes in the working world. However, it is up to sociology ensure the development of socially relevant subject: the subject which seeks autonomy, self-realization and emancipation, standing before the historical reality, thinking this reality and acting on it.

Philosophy

Science and philosophy have the same historical origins focused on rational explanation of natural phenomena, as opposed to mythological and religious arguments to justify.

Both are characterized by the intention to broaden the understanding of reality through the relentless pursuit of knowledge: philosophy, to grasp it in its entirety, and Science through an organized body of knowledge especially obtained by observation and experience.

Superficially can say that both are scientific knowledge, having as object the same reality, but distinguished by inexperimentável or experiential perspective adopted.

The consideration of epistemological issues such as the possibility of scientific knowledge, the conditions for the revelation of the true knowledge and the relationship between scientific theories and the experience for them portrayed are vital points in shaping the contemporary professional.

Thus the Computer comprehensive study transcends the purely technical issues, requiring a priori understanding of the knowledge building process, the central point of any philosophical investigation.

Methodology

Computer and Computer Area Courses can be divided into four broad categories, not equivalent to each other:

- courses that predominantly have computing as a core activity;
- courses that predominantly have computing as a means activity;
- the degree courses in Computer and

- Technology courses (community colleges)

1) Courses that have computing as a core activity aimed at training human resources for the scientific and technological development of computing. The graduates of these courses must be located in the state of the art science and computer technology, so that they can continue their activities in research, promoting scientific development, or applying scientific knowledge, promoting technological development. It should be given in these courses a strong emphasis on the use of laboratories to enable graduates in the design and construction of software and hardware design. The institution thirst for a course of this caliber should develop research activities in computing and students, it participating, will lead to the labor market and innovative ideas will have the ability to leverage and / or transform the labor market. So, are important human resources for the future market through entrepreneurial activities, the software industries and computers. The graduates of these courses are also potential candidates to pursue an academic career, through post-graduate studies. It is recommended that the courses in this category are developed at universities that have graduate in computing. A large portion of the teachers responsible for the computing disciplines must give full devotion to the institution with a view to research, extension and graduate. The curriculum of these courses should include a Working graduation (completion of course work) to be developed during a semester, which contributes to the technological development of computing. These courses, given its characteristics, preferably, should be developed in the morning or evening shifts. It is estimated that the market requires 25-50% of graduates of these courses on the total of graduates needed for the computing market. These courses are called BS in Computer Science or Computer Engineering.

The application of computer science and the use of computer technology in computer science courses are specific to each course.

There is no consensus on the profile difference between the so-called courses of Computer Science and Computer Engineering. Typically, the difference is in the application of science Computer and use of computing technology: Computer Engineering courses aimed at the application of computer science and the use of computer technology, specifically in solving problems related to industrial automation . Many Computer Engineering courses aim also the application of physics and electricity in solving the problems of industrial automation. These courses therefore include in their curricula, a new scientific basis, physics and electricity, which introduced a comprehensive and profound way too extend the curricula of the courses, in addition to invade the area of competence of the electrical engineering.

Automation - Automation area involves all the work of transformation activities originally played by humans in tasks performed by computer systems, in order to increase productivity, efficiency and safety, and cost reduction. Thus, an Automation System adds a set of equipment, information systems and procedures which is to automatically perform productive tasks with minimal human interference. The procedures implementing the processes which can be classified into three categories: Continuous Processes (continuous production flow, where the variables are analog, for example, in the chemical industry, steel, etc.); Manufacturing Processes (Discrete) (discrete production flow, originating from industrial application intensive labor such as, e.g. in the automotive industry); and Service Processes (where the final product is a service, such as in the case of the financial industry, commerce and engineering).

Industrial Automation - Industrial Automation refers to the first two types of the aforementioned methods (continuous or discrete).

Industrial Automation is a multidisciplinary field of technology and requires the integration of knowledge of basic areas, technological and even complementary, such as:

- Physics, Electricity and systems control, for the design of process controllers systems;
- Computer architecture, for the specification and design of systems that meet the functional requirements of the applications to be controlled, the project supervision and control interfaces (data acquisition and operation of the controlled environment);
- Real-time systems, the verification of the temporal aspects of the process, from requirements specification, through the specific features of operating systems and even architecture and communication processors that satisfy such conditions;
- Computer networks, mainly local, with its various configurations and communication protocols;

- Distributed systems, particularly for the software, synchronization, collaborative work;
- Software engineering, system design involving timing requirements;
- System reliability in environments with varying degrees of hostility, redundant architectures, hardware and software robustness;
- Other areas Computing: Neural Networks and Fuzzy Robotics systems like matter that can be seen as an industrial automation tool;

2) The courses that have computing as a means activity aimed at training human resources for automation of information systems organizations. Courses should give a strong emphasis on the use of laboratories to enable graduates "in use" efficient technologies in organizations. These courses bring together computer technology and management of technology and therefore have, in both areas, a strong and little theoretical pragmatic approach. It is very important that students perform internships in organizations and on the faculty have a good experience in the area of market information systems. They are important human resources to meet current labor market needs. The graduates of these courses should seek, when necessary, an update of their education through specialization courses (lato sensu graduate) and are potential candidates for stricto sensu graduate courses, responsible for the scientific development of the area of information systems organizations. The curriculum of these courses should include a Working graduation (completion of course work), to be developed during a semester, which contributes to the improvement of automation, performance, efficiency and rationalization of administrative services organizations. These courses, given their characteristics can also be developed on the night shift. It is recommended that the courses in this category are developed in university centers, integrated schools and colleges. It is estimated that the market requires 50-75% of graduates of these courses on the total of graduates needed for the computing market. These courses are called Bachelor of Information Systems.

Automation - Automation area involves all the work of transformation activities originally played by humans in tasks performed by computer systems, in order to increase productivity, efficiency and safety, and cost reduction. Thus, an Automation System adds a set of equipment, information systems and procedures which is to automatically perform productive tasks with minimal human interference. The procedures implementing the processes which can be classified into three categories: Continuous Processes (continuous production flow, where the variables are analog, for example, in the chemical industry, steel, etc.); Manufacturing Processes (Discrete) (discrete production flow, originating from industrial application intensive labor such as, e.g. in the automotive industry); and Service Processes (where the final product is a service, such as in the case of the financial industry, commerce and engineering).

Automation of Information Systems - Information Systems Automation refers to the third type of the above cases

Courses working information systems, in the academic field, covering two major areas: (1) the acquisition, development and management of services and resources of information technology and (2) the development and evolution of systems and infrastructure for use in organizational processes.

The information systems function has the overall responsibility to develop, implement and manage an information technology infrastructure (computers and communications) data (internal and external) and systems that span the entire organization. It is responsible for prospecting of new information technologies and assist in its incorporation into the strategies, planning and practices of the organization. The function also supports technology systems and individual departmental information.

The systems development activity to organizational and inter-organizational processes involves the creative use of information technology for data acquisition, communication, coordination, analysis and decision support. There are methods, techniques, technology and methodologies for this activity. The creation of systems in organizations includes issues of innovation, quality, human-machine systems, human-machine interfaces, socio-technical projects and change management.

Information systems are broadcast by all organizational functions. They are used for accounting, finance, sales, production and so on. This widespread use increases the need for professionals with knowledge of information systems development and systems management. Professionals with these skills support innovation, planning and information management infrastructure and coordination of information resources. The development of information by members of the SI team systems involves not only integrated systems encompassing the entire organization, but also support for the development of departmental and individual applications. "

Information systems can be defined as a combination of human and computer resources that interrelate the collection, storage, retrieval, distribution and use of data for the purpose of managerial efficiency (planning, control, communication and decision making), in organizations. In addition, information systems can also help managers and users to analyze problems, create new products and services and visualize complex issues. The Information Systems study and its development involves multiple perspectives and multidisciplinary knowledge that include various fields of knowledge such as computer science, behavioral science, decision science, management science, political science, operations research, sociology, accounting, etc. .

This view indicates that information systems are social systems composed of information technology requiring social, organizational and intellectual investments to make them work properly.

It is understood by information technology as a combination of hardware and general or specific use software, including information systems, combined with storage technologies, distribution, telecommunications and visualization through the various media and their techniques. With economic growth of information and the need for its global distribution, entire industries are being transformed through the application of information and communication technologies. At the organizational level, many companies rely on this technology to its key functions such as production and sales, existing today very few areas that were not affected by information technology.

Thus, information systems are best known for the benefits they bring to the business management in which it tries to eliminate waste, the overly repetitive tasks, with or without the use of paper, in order to improve cost control, quality the product or service, maximizing the benefits achieved with the use of information technology.

To improve management efficiency, information systems of organizations should be integrated and are designed to anticipate future uncertainties in a dynamic environment that includes, in addition to its users and developers, the relationship with other organizations as clients (for commercial purposes or social), suppliers, competitors, regulatory agencies, etc.

3) Degree courses in Computer aim to train teachers for secondary education in institutions that introduce computing in their curricula. The correct way to introduce computing in high school is still little known. It is recommended that the courses are developed in this category Institutes or Schools Professional high school may have in computing one of its alternatives, when professionals to meet specific needs of the area may be necessary.

4) Courses technology, under the law, are higher-level courses designed to meet emergency needs of the labor market and, therefore, are short-lived and terminals. Once satisfied the demand of professional courses must be extinguished. There are no rules for curriculum design. There must be consistency between curriculum and name of the course. Computing and information technology area, because it is dynamic, is in technology courses an efficient solution to address immediate and urgent needs of the labor market. Under current legislation they can be classified as sequential courses. It is recommended that the courses in this category are developed in university centers, integrated schools and colleges. The Technology courses in Data Processing, created in the 70s to replace the training of human resources for companies providing computers, should be extinguished / converted, since there is continuing need for training of human resources to meet this market segment. The full Bachelor courses in Information Systems replace the current technology courses in Data Processing with great advantages.

The following shows what subjects that should make each of computing and computer area courses profiles and how they should be detailed.

courses materials	Bachelor of Computer Science	Computer engineering	Bachelor of Information Systems	Degree in Computer
3.1.1.1 programming	The courses should cover, with breadth and depth, at least one programming language this matter (first programming language). They should also cover with breadth and depth paradigms of programming languages, data structure and search and sort data	The courses should cover, with breadth and depth, at least one programming language this matter (first programming language). They should also cover with breadth and depth paradigms of programming languages, data structure and search and sort data	The courses must cover all major programming languages with breadth and depth. They should also cover with breadth and depth data structure and search and sort data	The courses must cover all major programming languages with breadth and depth. They should also cover with breadth and depth data structure and search and sort data.
3.1.1.2 Computation and Algorithms	The courses should cover this matter with breadth and depth	The courses should cover this matter with breadth and depth	The courses should cover this issue in a comprehensive and general	The courses should cover this matter with breadth and depth
3.1.1.3 Computer Architecture	The courses should cover this matter with breadth and depth	The courses should cover this matter with breadth and depth	The courses should cover this issue in a comprehensive and general.	The courses should cover this issue in a comprehensive and general.
3.1.2 Mathematics	The courses should cover discrete mathematics, graph theory, combinatorics and logic of this matter in breadth and depth. The remaining contents of this matter should be covered according to the degree of coverage and depth with which the materials of technological training are introduced and the types of problems to be solved with mathematics (statistics, operational research etc.)	The courses should cover the content of discrete mathematics, graph theory, combinatorics and logic of this matter in breadth and depth. The remaining contents of this matter should be covered according to the degree of coverage and depth with which the materials of technological training are introduced and the types of problems to be solved with mathematics (statistics, operational research etc.)	The courses should cover discrete mathematics and logic of this matter in a comprehensive and general. The remaining contents of this matter should be covered according to the degree of coverage and depth with which the materials of additional training are introduced and the types of problems to be solved with mathematics (statistics, operational research etc.)	The courses should cover discrete mathematics, graphs, combinatorial analysis and logic of this matter in breadth and depth.

3.1.3 Physics and Electricity	The courses should cover this subject in breadth and depth enough that students understand the physical implementation of logical devices and can perform hardware designs. Students should, in the laboratory, conducting experiments, such as the assembly of simple logic circuits, observing the electrical phenomena involved in the interaction of the components, observe the phenomena involved in data communication, and simulate more complex systems and processor architectures and system models most complex computational, as communications equipment, networks and algorithms used in operating systems.	The courses should cover this subject in breadth and depth enough that students understand the physical implementation of logical devices and can perform hardware designs. Students should, in the laboratory, conducting experiments, such as the assembly of simple logic circuits, observing the electrical phenomena involved in the interaction of the components, observe the phenomena involved in data communication, and simulate more complex systems and processor architectures and system models most complex computational, as communications equipment, networks and algorithms used in operating systems.	This material is dispensable	This material is dispensable
3.1.4 Pedagogy	This material is dispensable	This material is dispensable	This material is dispensable	The courses should cover this matter with breadth and depth

3.2 Technological Training	The courses should cover the foundations / structures of all technologies and at least one of them (emphasis) in depth with a view to carrying out projects. The Practice of Computer Teaching to be missed.	The courses should cover the foundations / structures of all technologies and at least one of them (emphasis) in depth with a view to carrying out projects. The Practice of Computer Teaching to be missed.	The courses should cover the foundations / structures of all technologies, without, however, to enable students to design them. Should be used extensively in the laboratory, current technology: database, software engineering, computer networks, among others. The Practice of Computer Teaching to be missed.	The courses should cover the foundations / structures of all technologies, without, however, to enable students to design them. The courses should cover the subject "Practical Computer education" with breadth and depth, totaling this coverage, in hours, as determined by LDB (Art. 65), and the application of focused pedagogy for "how to teach in general."
3.3 additional training areas.	The courses must cover at least one other area of human resources training, so that the course graduates to interact with their own professionals in solving their problems. In addition, graduates must understand, in general, the problems that affect them as professionals: economics, business, law, among others.	The courses should cover the system control areas and system reliability. In addition, graduates must understand, in general, the problems that affect them as professionals: economics, business, law, among others.	The courses should cover, among others, behavioral science, decision science, management science, political science, operations research, sociology, economics, accounting and general theory of such systems that the course graduates can understand in depth the functions of problems organizations, planning, control, communication, decision making, accounting, finance, sales, production, according to the profile of the course described above.	This matter is expendable.

3.4 Formation humanistic	The courses should cover this subject in general.	The courses should cover this subject in general.	The courses should cover this subject in general.	The courses should cover this subject in general.

Note: A general education in some matter, contrary to a deep formation is obtained by taking knowledge of the matter succinctly.

MINIMUM TIMES FOR THE COURSES OF COMPUTER AND INFORMATION TECHNOLOGY AREA

It is recommended that the upper courses of computing and information technology area have the registration scheme semester course or biannual series regime. Each semester will have at least 400 hours of effective academic work, distributed at least 100 working days, excluding the time reserved for final exams, if any. Technology courses must take four semesters and graduate courses, at least eight semesters.

4 MACHINE LEARNING APPLICATIONS IN CANCER PROGNOSIS AND PREDICTION

Konstantina Kourou[a], Themis P. Exarchos[a,b], Konstantinos P. Exarchos[a], Michalis V. Karamouzis[c], Dimitrios I. Fotiadis[a,b]

[a]Unit of Medical Technology and Intelligent Information Systems, Dept. of Materials Science and Engineering, University of Ioannina, Ioannina, Greece

[b]IMBB — FORTH, Dept. of Biomedical Research, Ioannina, Greece

[c]Molecular Oncology Unit, Department of Biological Chemistry, Medical School, University of Athens, Athens, Greece

Keywords: Machine learning; Cancer susceptibility; Predictive models; Cancer recurrence; Cancer survival

ABSTRACT

Cancer has been characterized as a heterogeneous disease consisting of many different subtypes. The early diagnosis and prognosis of a cancer type have become a necessity in cancer research, as it can facilitate the subsequent clinical management of patients. The importance of classifying cancer patients into high or low risk groups has led many research teams, from the biomedical and the bioinformatics field, to study the application of machine learning (ML) methods. Therefore, these techniques have been utilized as an aim to model the progression and treatment of cancerous conditions. In addition, the ability of ML tools to detect key features from complex datasets reveals their importance. A variety of these techniques, including Artificial Neural Networks (ANNs), Bayesian Networks (BNs), Support Vector Machines (SVMs) and Decision Trees (DTs) have been widely applied in cancer research for the development of predictive models, resulting in effective and accurate decision making. Even though it is evident that the use of ML methods can improve our understanding of cancer progression, an appropriate level of validation is needed in order for these methods to be considered in the everyday clinical practice. In this work, we present a review of recent ML approaches employed in the modeling of cancer progression. The predictive models discussed here are based on various supervised ML techniques as well as on different input features and data samples. Given the growing trend on the application of ML methods in cancer research, we present here the most recent publications that employ these techniques as an aim to model cancer risk or patient outcomes.

INTRODUCTION

Over the past decades, a continuous evolution related to cancer research has been performed [1]. Scientists applied different methods, such as screening in early stage, in order to find types of cancer before they cause symptoms. Moreover, they have developed new strategies for the early prediction of cancer treatment outcome. With the advent of new technologies in the field of medicine, large amounts of cancer data have been collected and are available to the medical research community. However, the accurate prediction of a disease outcome is one of the most interesting and challenging tasks for physicians. As a result, ML methods have become a popular tool for medical researchers. These techniques can discover and identify patterns and relationships between them, from complex datasets, while they are able to effectively predict future outcomes of a cancer type.

Given the significance of personalized medicine and the growing trend on the application of ML techniques, we here present a review of studies that make use of these methods regarding the cancer prediction and prognosis. In these studies prognostic and predictive features are considered which may be independent of a certain treatment or are integrated in order to guide therapy for cancer patients, respectively [2]. In addition, we discuss the types of ML methods being used, the types of data they integrate, the overall performance of each proposed scheme while we also discuss their pros and cons.

An obvious trend in the proposed works includes the integration of mixed data, such as clinical and genomic. However, a common problem that we noticed in several works is the lack of external validation or testing regarding the predictive performance of their models. It is clear that the application of ML methods could improve the accuracy of cancer susceptibility, recurrence and survival prediction. Based on [3], the accuracy of cancer prediction outcome has significantly improved by 15%–20% the last years, with the application of ML techniques.

Several studies have been reported in the literature and are based on different strategies that could enable the early cancer diagnosis and prognosis [4]; [5]; [6] ; [7]. Specifically, these studies describe approaches related to the profiling of circulating miRNAs that have been proven a promising class for cancer detection and identification. However, these methods suffer from low sensitivity regarding their use in screening at early stages and their difficulty to discriminate benign from malignant tumors. Various aspects regarding the prediction of cancer outcome based on gene expression signatures are discussed in [8] ; [9]. These studies list the potential as well as the limitations of microarrays for the prediction of cancer outcome. Even though gene signatures could significantly improve our ability for prognosis in cancer patients, poor progress has been made for their application in the clinics. However, before gene expression profiling can be used in clinical practice, studies with larger data samples and more adequate validation are needed.

In the present work only studies that employed ML techniques for modeling cancer diagnosis and prognosis are presented.

ML TECHNIQUES

ML, a branch of Artificial Intelligence, relates the problem of learning from data samples to the general concept of inference [10]; [11] ; [12].

Every learning process consists of two phases: (i) estimation of unknown dependencies in a system from a given dataset and (ii) use of estimated dependencies to predict new outputs of the system. ML has also been proven an interesting area in biomedical research with many applications, where an acceptable generalization is obtained by searching through an *n-dimensional* space for a given set of biological samples, using different techniques and algorithms [13]. There are two main common types of ML methods known as (i) supervised learning and (ii) unsupervised learning. In supervised learning a labeled set of training data is used to estimate or map the input data to the desired output. In contrast, under the unsupervised learning methods no labeled examples are provided and there is no notion of the output during the learning process. As a result, it is up to the learning scheme/model to find patterns or discover the groups of the input data. In supervised learning this procedure can be thought as a classification problem. The task of classification refers to a learning process that categorizes the data into a set of finite classes. Two other common ML tasks are regression and clustering. In the case of regression problems, a learning function maps the data into a real-value variable. Subsequently, for each new sample the value of a predictive variable can be estimated, based on this process. Clustering is a common unsupervised task in which one tries to find the categories or clusters in order to describe the data items. Based on this process each new sample can be assigned to one of the identified clusters concerning the similar characteristics that they share.

Suppose for example that we have collected medical records relevant to breast cancer and we try to predict if a tumor is malignant or benign based on its size. The ML question would be referred to the estimation of the probability that the tumor is malignant or no (1 = Yes, 0 = No). Fig. 1 depicts the classification process of a tumor being malignant or not. The circled records depict any misclassification of the type of a tumor produced by the procedure.

Another type of ML methods that have been widely applied is semi-supervised learning, which is a combination of supervised and unsupervised learning. It combines labeled and unlabeled data in order to construct an accurate learning model. Usually, this type of learning is used when there are more unlabeled datasets than labeled.

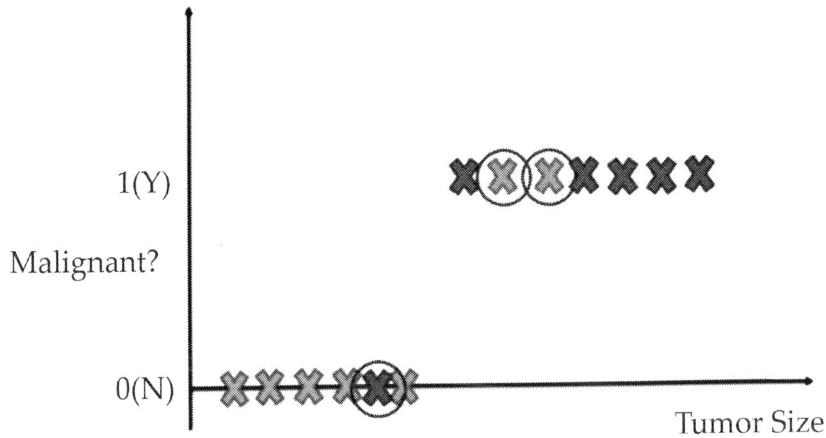

Figure 1: Classification task in supervised learning. Tumors are represented as X and classified as benign or malignant. The circled examples depict those tumors that have been misclassified.

When applying a ML method, data samples constitute the basic components. Every sample is described with several features and every feature consists of different types of values. Furthermore, knowing in advance the specific type of data being used allows the right selection of tools and techniques that can be used for their analysis. Some data-related issues refer to the quality of the data and the preprocessing steps to make them more suitable for ML. Data quality issues include the presence of noise, outliers, missing or duplicate data and data that is biased-unrepresentative. When improving the data quality, typically the quality of the resulting analysis is also improved. In addition, in order to make the raw data more suitable for further analysis, preprocessing steps should be applied that focus on the modification of the data. A number of different techniques and strategies exist, relevant to data preprocessing that focus on modifying the data for better fitting in a specific ML method. Among these techniques some of the most important approaches include (i) dimensionality reduction (ii) feature selection and (iii) feature extraction. There are many benefits regarding the dimensionality reduction when the datasets have a large number of features. ML algorithms work better when the dimensionality is lower [14]. Additionally, the reduction of dimensionality can eliminate irrelevant features, reduce noise and can produce more robust learning models due to the involvement of fewer features. In general, the dimensionality reduction by selecting new features which are a subset of the old ones is known as feature selection. Three main approaches exist for feature selection namely embedded, filter and wrapper approaches [14]. In the case of feature extraction, a new set of features can be created from the initial set that captures all the significant information in a dataset. The creation of new sets of features allows for gathering the described benefits of dimensionality reduction.

However, the application of feature selection techniques may result in specific fluctuations concerning the creation of predictive feature lists. Several studies in the literature discuss the phenomenon of lack of agreement between the predictive gene lists discovered by different groups, the need of thousands of samples in order to achieve the desired outcomes, the lack of biological interpretation of predictive signatures and the dangers of information leak recorded in published studies [15]; [16]; [17] ; [18].

The main objective of ML techniques is to produce a model which can be used to perform classification, prediction, estimation or any other similar task. The most common task in learning process is classification. As mentioned previously, this learning function classifies the data item into one of several predefined classes. When a classification model is developed, by means of ML techniques, training and generalization errors can be produced. The former refers to misclassification errors on the training data while the latter on the expected errors on testing data. A good classification model should fit the training set well and accurately classify all the instances. If the test error rates of a model begin to increase even though the training error rates decrease then the phenomenon of model overfitting occurs. This situation is related to model complexity meaning that the training errors of a model can be reduced if the model complexity increases. Obviously, the ideal complexity of a model not susceptible to overfitting is the one that produces the lowest generalization error. A formal method for analyzing the expected generalization error of a learning algorithm is the bias–variance decomposition. The bias component of a particular learning algorithm measures the error rate of that algorithm. Additionally, a second source of error over all possible training sets of given size and all possible test sets is called variance of the learning method. The overall expected error of a classification model is constituted of the sum of bias and variance, namely the bias–variance decomposition.

Once a classification model is obtained using one or more ML techniques, it is important to estimate the classifier's performance. The performance analysis of each proposed model is measured in terms of sensitivity, specificity, accuracy and area under the curve (AUC). Sensitivity is defined as the proportion of true positives that are correctly observed by the classifier, whereas specificity is given by the proportion of true negatives that are correctly identified. The quantitative metrics of accuracy and AUC are used for assessing the overall performance of a classifier. Specifically, accuracy is a measure related to the total number of correct predictions. On the contrary, AUC is a measure of the model's performance which is based on the ROC curve that plots the tradeoffs between sensitivity and 1-specificity (Fig. 2).

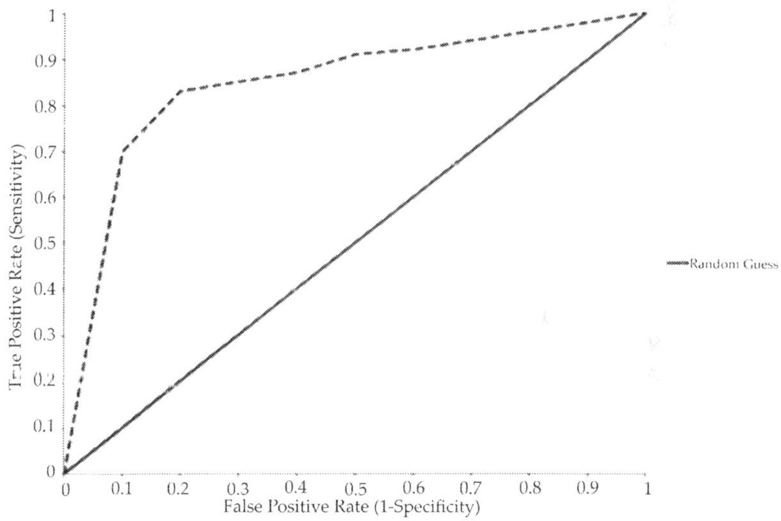

Figure 2: An indicative ROC curve of two classifiers: (a) Random Guess classifier (red curve) and (b) A classifier providing more robust predictions (blue dotted curve).

The predictive accuracy of the model is computed from the testing set which provides an estimation of the generalization errors. In order to obtain reliable results regarding the predicting performance of a model, training and testing samples should be sufficiently large and independent while the labels of the testing sets should be known. Among the most commonly used methods for evaluating the performance of a classifier by splitting the initial labeled data into subsets are: (i) Holdout Method, (ii) Random Sampling, (iii) Cross-Validation and (iv) Bootstrap. In the Holdout method, the data samples are partitioned into two separate sets, namely the training and the test sets. A classification model is then generated from the training set while its performance is estimated on the test set. Random sampling is a similar approach to the Holdout method. In this case, in order to better estimate the accuracy, the Holdout method is repeated several times, choosing the training and test instances randomly. In the third approach, namely cross-validation, each sample is used the same number of times for training and only once for testing. As a result, the original data set is covered successfully both in the training and in the test set. The accuracy results are calculated as the average of all different validation cycles. In the last approach, bootstrap, the samples are separated with replacement into training and test sets, i.e. they are placed again into the entire data set after they have been chosen for training.

When the data are preprocessed and we have defined the kind of learning task, a list of ML methods including (i) ANNs, (ii) DTs, (iii) SVMs and (iv) BNs is available. Based on the intension of this review paper, we will refer only to these ML techniques that have been applied widely in the literature for the case study of cancer prediction and prognosis. We identify the trends regarding the types of ML methods that are used, the types of data that are integrated as well as the evaluation methods employed for assessing the overall performance of the methods used for cancer prediction or disease outcomes.

ANNs handle a variety of classification or pattern recognition problems. They are trained to generate an output as a combination between the input variables. Multiple hidden layers that represent the neural connections mathematically are typically used for this process. Even though ANNs serve as a gold standard method in several classification tasks [19] they suffer from certain drawbacks. Their generic layered structure proves to be time-consuming while it can lead to very poor performance. Additionally, this specific technique is characterized as a "black-box" technology. Trying to find out how it performs the classification process or why an ANN did not work is almost impossible to detect. Fig. 3depicts the structure of an ANN with its interconnected group of nodes.

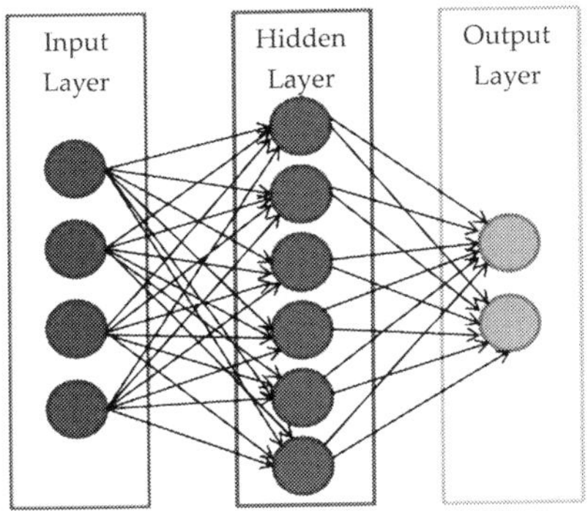

Figure 3: An illustration of the ANN structure. The arrows connect the output of one node to the input of another.

DTs follow a tree-structured classification scheme where the nodes represent the input variables and the leaves correspond to decision outcomes. DTs are one of the earliest and most prominent ML methods that have been widely applied for classification purposes. Based on the architecture of the DTs, they are simple to interpret and "quick" to learn. When traversing the tree for the classification of a new sample we are able to conjecture about its class. The decisions resulted from their specific architecture allow for adequate reasoning which makes them an appealing technique. Fig. 4 depicts an illustration of a DT with its elements and rules.

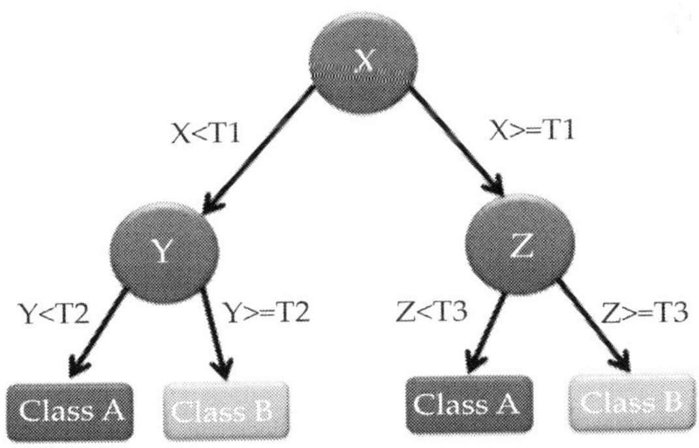

Figure 4: An illustration of a DT showing the tree structure. Each variable (X, Y, Z) is represented by a circle and the decision outcomes by squares (Class A, Class B). T(1–3) represents the thresholds (classification rules) in order to successfully classify each variable to a class label.

SVMs are a more recent approach of ML methods applied in the field of cancer prediction/prognosis. Initially SVMs map the input vector into a feature space of higher dimensionality and identify the hyperplane that separates the data points into two classes. The marginal distance between the decision hyperplane and the instances that are closest to boundary is maximized. The resulting classifier achieves considerable generalizability and can therefore be used for the reliable classification of new samples. It is worth noting that probabilistic outputs can also be obtained for SVMs [20]. Fig. 5 illustrates how an SVM might work in order to classify tumors among benign and malignant based on their size and patients' age. The identified hyperplane can be thought as a decision boundary between the two clusters. Obviously, the existence of a decision boundary allows for the detection of any misclassification produced by the method.

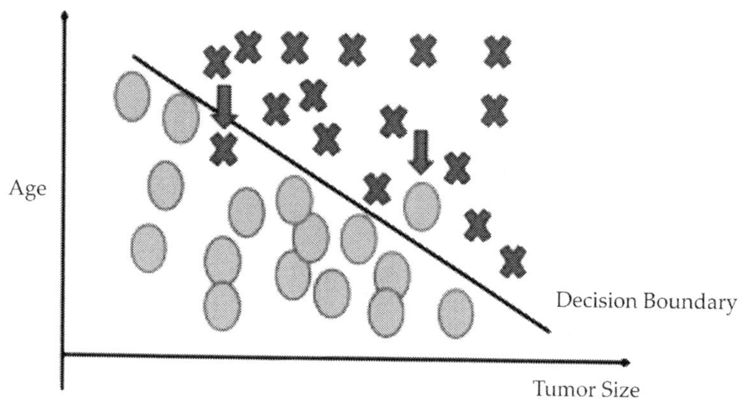

Figure 5: A simplified illustration of a linear SVM classification of the input data. Figure was reproduced from the ML lectures of [21]. Tumors are classified according to their size and the patient's age. The depicted arrows display the misclassified tumors.

BN classifiers produce probability estimations rather than predictions. As their name reveals, they are used to represent knowledge coupled with probabilistic dependencies among the variables of interest via a directed acyclic graph. BNs have been applied widely to several classification tasks as well as for knowledge representation and reasoning purposes.

Fig. 6 depicts an illustration of a BN across with the calculated conditional probability for each variable.

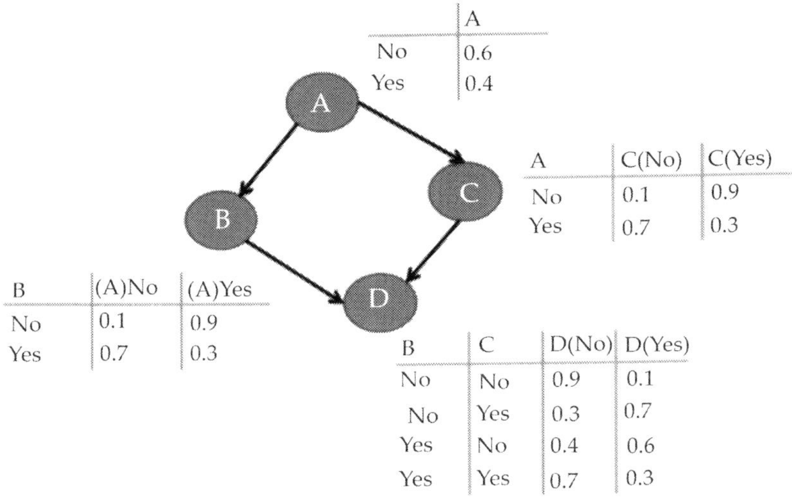

Figure 6: An illustration of a BN. Nodes (A–D) represent a set of random variables across with their conditional probabilities which are calculated in each table.

ML AND CANCER PREDICTION/PROGNOSIS

The last two decades a variety of different ML techniques and feature selection algorithms have been widely applied to disease prognosis and prediction [3]; [22]; [23]; [24]; [25]; [26] ; [27]. Most of these works employ ML methods for modeling the progression of cancer and identify informative factors that are utilized afterwards in a classification scheme. Furthermore, in almost all the studies gene expression profiles, clinical variables as well as histological parameters are encompassed in a complementary manner in order to be fed as input to the prognostic procedure. Fig. 7 depicts the distribution in published papers using ML techniques to predict (i) cancer susceptibility, (ii) recurrence and (iii) survival. The information was collected based on a variety of query searches in the Scopus biomedical database. More specifically, queries like "cancer risk assessment" AND "Machine Learning", "cancer recurrence" AND "Machine Learning", "cancer survival" AND "Machine Learning" as well as "cancer prediction" AND "Machine Learning" yielded the number of papers that are depicted in Fig. 3. No limitations were imposed in the resulted hits except the exclusion of articles published before 2010. As mentioned above, the number of papers presented in Fig. 7 refers to the exact numbers yielded from the databases without any refinement except the date that they were published.

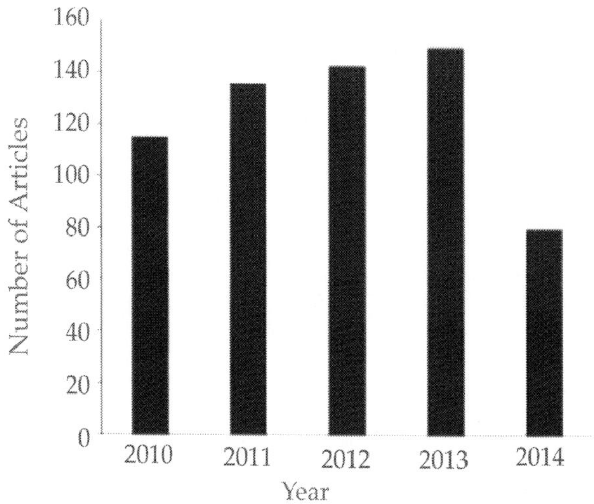

Figure 7: Distribution of published studies, within the last 5 years, that employ ML techniques for cancer prediction.

The success of a disease prognosis is undoubtedly dependent on the quality of a medical diagnosis; however, a prognostic prediction should take into account more than a simple diagnostic decision. When dealing with cancer prognosis/prediction one is concerned with three predictive tasks: (i) the prediction of cancer susceptibility (risk assessment), (ii) the prediction of cancer recurrence/local control and (iii) the prediction of cancer survival. In the first two cases one is trying to find (i) the likelihood of developing a type of cancer and (ii) the likelihood of redeveloping a type of cancer after complete or partial remission. In the last case, the prediction of a survival outcome such as disease-specific or overall survival after cancer diagnosis or treatment is the main objective. The prediction of cancer outcome usually refers to the cases of (i) life expectancy, (ii) survivability, (iii) progression and (iv) treatment sensitivity [3].

Major types of ML techniques including ANNs and DTs have been used for nearly three decades in cancer detection [22]; [28]; [29] ; [30]. According to the recent PubMed results regarding the subject of ML and cancer more than 7510 articles have been published until today. The vast majority of these publications makes use of one or more ML algorithms and integrates data from heterogeneous sources for the detection of tumors as well as for the prediction/prognosis of a cancer type. A growing trend is noted the last decade in the use of other supervised learning techniques, namely SVMs and BNs, towards cancer prediction and prognosis [24]; [31]; [32]; [33]; [34]; [35] ; [36]. All of these classification algorithms have been widely used in a wide range of problems posed in cancer research.

In the past, the typical information used by the physicians conclude with a reasonable decision regarding cancer prognosis and included histological, clinical and population-based data [23] ; [37]. The integration of features such as family history, age, diet, weight, high-risk habits and exposure to environmental carcinogens play a critical role in predicting the development of cancer [38]; [39] ; [40]. Even though this type of macro-scale information referred to a small number of variables so that standard statistical methods could be used for prediction purposes, however these types of parameters do not provide sufficient information for making robust decisions. With the rapid advent of genomic, proteomic and imaging technologies a new kind of molecular information can be obtained. Molecular biomarkers, cellular parameters as well as the expression of certain genes have been proven as very informative indicators for cancer prediction. The presence of such High Throughput Technologies (HTTs) nowadays has produced huge amounts of cancer data that are collected and are available to the medical research community. However, the accurate prediction of a disease outcome is one of the most interesting and challenging tasks for physicians. As a result, ML methods have become a popular tool for medical researchers. These techniques can discover and identify patterns and relationships between them, from complex datasets, while they are able to effectively predict future outcomes of a cancer type. Additionally, feature selection methods have been published in the literature with their application in cancer [41]; [42] ; [43]. The proposed computational tools aim at identifying informative features for accurately identification of disease class.

There are nowadays separate subgroups among the same type of cancer based on specific genetic defects that have different treatment approaches and options as well as different clinical outcomes. This is the foundation of the individualized treatment approach, in which computational techniques could help by identifying less costly and effectively such small groups of patients. Furthermore, the development of a community resource project, namely The Cancer Genome Atlas Research Network (TCGA) has the potential support for personal medicine as it provides large scale genomic data about specific tumor types. TCGA provides with the ability to better understand the molecular basis of cancer through the application of high-throughput genome technologies.

SURVEY OF ML APPLICATIONS IN CANCER

An extensive search was conducted relevant to the use of ML techniques in cancer susceptibility, recurrence and survivability prediction. Two electronic databases were accessed namely PubMed, Scopus. Due to the vast number of articles returned by the search queries, further scrutinization was needed in order to maintain the most relevant articles. The relevance of each publication was assessed based on the keywords of the three predictive tasks found in their titles and abstracts. Specifically, after reading their titles and abstracts we only selected those publications that study one of the three foci of cancer prediction and included it in their titles. The majority of these studies use different types of input data: genomic, clinical, histological, imaging, demographic, epidemiological data or combination of these. Papers that focus on the prediction of cancer development by means of conventional statistical methods (e.g. chi-square, Cox regression) were excluded as were papers that use techniques for tumor classification or identification of predictive factors. According to [3] and their survey based on ML applications in cancer prediction, we noted a rapid increase in papers that have been published in the last decade. Although it is impossible to achieve a complete coverage of the literature, we believe that a significant number of relevant papers were extracted and are presented in this review. As mentioned above, from the initial group of papers we selected a representative list that follows a well-organized structure. Specifically, we selected these studies that make use of recognizable ML techniques and integrated data from heterogeneous sources in order to predict the desirable outcome. We focused mainly on studies that have been published the last 5 years as an aim to present the most recent state of the art in the field and their advances in comparison to older publications. Table 1a, Table 1b ; Table 1c depict some of the publications presented in this review. Cancer type, ML method, number of patients, type of data as well as the overall accuracy achieved by each proposed method are presented. Each sub-table corresponds to studies regarding a specific scenario (i.e. cancer susceptibility prediction, cancer recurrence prediction and cancer survival prediction). It should be noted that in articles that more than one ML techniques are applied for prediction, we decided to present here the most accurate predictive model.

Table 1a: Publications relevant to ML methods used for cancer susceptibility prediction.

Publication	Method	Cancer type	No of patients	Type of data	Accuracy	Validation method	Important features
Ayer T et al. [19]	ANN	Breast cancer	62,219	Mammographic, demographic	AUC = 0.965	10-fold cross validation	Age, mammography findings
Waddell M et al. [44]	SVM	Multiple myeloma	80	SNPs	71%	Leave-one-out cross validation	snp739514, snp521522, snp994532
Listgarten J et al. [45]	SVM	Breast cancer	174	SNPs	69%	20-fold cross validation	snpCY11B2 (+) 4536 T/C snpCYP1B1 (+) 4328 C/G
Stajadinovic et al. [46]	BN	Colon carcinomatosis	53	Clinical, pathologic	AUC = 0.71	Cross-validation	Primary tumor histology, nodal staging, extent of peritoneal cancer

Table 1b: Publications relevant to ML methods used for cancer recurrence prediction.

Publication	ML method	Cancer type	No of patients	Type of data	Accuracy	Validation method	Important features
Exarchos K et al.[24]	BN	Oral cancer	86	Clinical, imaging tissue genomic, blood genomic	100%	10-fold cross validation	Smoker, p53 stain, extra-tumor spreading, TCAM, SOD2
Kim W et al. [47]	SVM	Breast cancer	679	Clinical, pathologic, epidemiologic	89%	Hold-out	Local invasion of tumor
Park C et al. [48]	Graph-based SSL algorithm	Colon cancer, breast cancer	437 374	Gene expression, PPIs	76.7% 80.7%	10-fold cross validation	BRCA1, CCND1, STAT1, CCNB1
Tseng C-J et al. [49]	SVM	Cervical cancer	168	Clinical, pathologic	68%	Hold-out	pathologic_S, pathologic_T, cell type RT target summary
Eshlaghy A et al.[34]	SVM	Breast cancer	547	Clinical, population	95%	10-fold cross validation	Age at diagnosis, age at menarche

Table 1c: Publications relevant to ML methods used for cancer survival prediction.

Publication	ML method	Cancer type	No of patients	Type of data	Accuracy	Validation method	Important features
Chen Y-C et al. [50]	ANN	Lung cancer	440	Clinical, gene expression	83.5%	Cross validation	Sex, age, T_stage, N_stage LCK and ERBB2 genes
Park K et al. [26]	Graph-based SSL algorithm	Breast cancer	162,500	SEER	71%	5-fold cross validation	Tumor size, age at diagnosis, number of nodes
Chang S-W et al.[32]	SVM	Oral cancer	31	Clinical, genomic	75%	Cross validation	Drink, invasion, p63 gene
Xu X et al.[51]	SVM	Breast cancer	295	Genomic	97%	Leave-one-out cross validation	50-gene signature

Gevaert O et al. [52]	BN	Breast cancer	97	Clinical, microarray	AUC = 0.851	Hold-Out	Age, angioinvasion, grade MMP9, HRASLA and RAB27B genes
Rosado P et al. [53]	SVM	Oral cancer	69	Clinical, molecular	98%	Cross validation	TNM_stage, number of recurrences
Delen D et al. [54]	DT	Breast cancer	200,000	SEER	93%	Cross validation	Age at diagnosis, tumor size, number of nodes, histology
Kim J et al. [36]	SSL Co-training algorithm	Breast cancer	162,500	SEER	76%	5-fold cross validation	Age at diagnosis, tumor size, number of nodes, extension of tumor

A detailed analysis of more recent studies revealed that there is a growing trend in risk assessment as well as the prediction of recurrence of a cancer type regardless the ML technique used. Many research groups have tried to predict the possibility of redeveloping cancer after remission and appeared to improve the accuracy of predictions compared to alternative statistical techniques. Moreover, the vast majority of these publications used molecular and clinical data in order to make their predictions. The use of such measurable features as input data is a growing trend based on the advent of HTTs.

In the following, we are going to discuss one case for each of the objectives of predicting (i) susceptibility, (ii) recurrence and (iii) survival, all by means of ML techniques. Each sub-section summarizes the representative studies we have selected based on their predictive outcomes. We only selected those publications that have been accepted the last 5 years and make use of distinguishable ML methods. We provide the readers with the appropriate details of the most recent techniques used for the prediction and prognosis of most frequent cancer types.

Prediction of Cancer Susceptibility

We performed a Scopus and a PubMed advanced search which was limited to the last 5 years. Out of these results one of the publications employs ML techniques for the prediction of susceptibility in a cancer type [55]. The authors perform a genetic epidemiology study of bladder cancer susceptibility in terms of Learning Classifying Systems (LCSs). We decided to exclude this work from the present case study as it deals with genetic information and examines further genetic problems. Based on these limitations we continued our search to the specific biomedical databases. Most of these titles neither referred to the specified keywords that are mentioned in the relevant survey nor used ML techniques for their predictions. Among the most recent publications that resulted after our limited literature search regarding the cancer risk assessment prediction [19]; [56]; [57] ; [58], we selected a recent and very interesting study to present relevant to the breast cancer risk estimation by means of ANNs [19]. It is a different study among the others presented in this review article regarding the data type used. Although all of the publications selected make use of molecular, clinical or population-based data, this work encompasses mammographic findings and demographic characteristics to the model. Even though this work doesn't fit our general statement regarding our search criteria, we decided to include it in this case study because no other search result met our needs. We excluded this work from our general statement because no other search result met our needs. The major intense in developing decision-making tools that can discriminate among benign and malignant findings in breast cancer is commented by the authors. They also mention that when developing prediction models, risk stratification is of major interest. According to their knowledge, existing studies based on the use of computer models, have also utilized specific ML techniques, such as ANNs, in order to assess the risk of breast cancer patients. In their work, ANNs are employed in order to develop a prediction model that could classify malignant mammographic findings from benign. They built their model with a large number of hidden layers which generalizes better than networks with small number of hidden nodes. Regarding the collected data in this study, 48.774 mammographic findings as well as demographic risks factors and tumor characteristics were considered. All of the mammographic records were reviewed by radiologists and the reading information was obtained. This dataset was then fed as input to the ANN model. Its performance was estimated by means of ten-fold cross validation. Additionally, in order to prevent the case of overfitting the authors used the ES approach. This procedure, generally, controls the network error during training and stops it if overfitting occurs. The calculated AUC of their model was 0.965 following training and testing by means of ten-fold cross validation. The authors claimed that their model can accurately estimate the risk assessment of breast cancer patients by integrating a large data sample. They also declared that their model is unique among others if we consider that the most important factors they used to train the ANN model are the mammography findings with tumor registry outcomes. One very interesting characteristic in this study is the calculation of two main components of accuracy, namely discrimination and calibration. Discrimination is a metric that someone calculates in order to separate benign abnormalities from malignant ones, while calibration is a measurement used when a risk prediction model aims to stratify patients into high or low risk categories. The authors plotted (i) a ROC curve in order to evaluate the discriminative ability of their model and (ii) a calibration curve for comparing afterwards their model's calibration to the perfect calibration of predicting breast cancer risk. Apart from these findings, the authors also noted that the use of a mix of screening and diagnostic datasets cannot be reliably separated when feeding as input to the ANN. So, in order to overcome such limitations the authors should consider the purpose of preprocessing steps for transforming the raw data into appropriate formats for subsequent analysis.

Prediction of Cancer Recurrence

Based on our survey, we here present the most relevant and recent publications that proposed the use of ML techniques for cancer recurrence prediction. A work which studies the recurrence prediction of oral squamous cell carcinoma (OSCC) is proposed in [24]. They suggested a multiparametric Decision Support System in order to analyze the basis of OSCC evolvement after total remission of cancer patients. They exploited heterogeneous sources of data (clinical, imaging and genomic) in order to predict a possible relapse of OSCC and thus a subsequent recurrence. A total number of 86 patients were considered in this study, 13 of which have been identified with a relapse while the remaining was disease free. A specific feature selection procedure was followed with the employment of two feature selection algorithms, namely CFS [59] and wrapper algorithm [60]. As a result, any bias could be avoided when selecting the most informative features of their reference heterogeneous dataset. Then the selected important variables could be used as input vectors to specific classifiers. Before the employment of the feature selection techniques the total number of the clinical, imaging and genomic features was 65, 17 and 40 in each category. Subsequently, after the employment of the CFS algorithm the total number of clinical, imaging and genomic data used in each classifier was 8, 6 and 7, respectively. More specifically, among the clinical variables the most informative ones, for each classification algorithm, were the smoker, tumor thickness and p53 stain. Concerning the imaging and the genomic features, after the utilization of the CFS algorithm, the most important were the extra-tumor spreading, the number of lymph nodes and the SOD2, TCAM and OXCT2 genes.

The basic idea in this study is summarized in the discrimination of patients into those with a disease relapse and those without after the performance of five classification algorithms. The employed algorithms include the BNs, ANNs, SVMs, DTs and RF classifiers. After the performance of each ML method an evaluation technique, namely ten-fold cross-validation, was employed for evaluation purposes. Additionally, accuracy, sensitivity and specificity were also calculated for comparison reasons among the employed classification schemes. The analysis of ROC curve was considered by the authors for evaluation purposes as well. Their predictive results regarding the classification schemes employed were obtained based on the classification of data without performing feature selection and on the classification of data after employing a feature selection algorithm. Regarding their outputs the authors claimed that the BN classifier without applying any feature selection scheme performed better in the discrimination with directly input of the clinical and imaging features (78.6% and 82.8% accuracy, respectively). In a similar manner, genomic-based classification results revealed that the best performing classifier was the BN in conjunction with the CFS algorithm (91.7% accuracy). In the final stage of their study, the authors combined the more accurate individual predictors (i.e. BN and BN coupled with the CFS) in order to yield a consensus decision for discrimination between patients with and without an OSCC relapse. A comparison of this approach to other studies in the literature revealed that this proposal yields robust results than other methodologies. The proposed study illustrated in an explanatory way how the integration of heterogeneous sources of data, by means of ML classifiers, can produce accurate results regarding the prediction of cancer recurrence. Furthermore, the authors used more than one classification technique in order to obtain robust results. It is clear that when you estimate the performance of a classifier predictor among others, then you are able to find the most optimal tool. However, we should highlight an important aspect of this work regarding the small sample size. Only 86 patients were considered with their clinical, imaging and genomic features. Although their classification results were very promising, we should consider that a relatively small sample size compared to data dimensionality can lead to misclassification and biased predictors. Another interesting article published in the same year with [24] proposed an SVM-based model for the prediction of breast cancer recurrence, called BCRSVM [47]. The authors support the idea that the classification of cancer patients into high-risk or low-risk groups allows experts to adjust a better treatment and follow-up planning. Their study is based on the development of a predictive model regarding the breast cancer recurrence within five years after surgery. SVM, ANN as well as Cox-proportional hazard regression were employed for producing the models and find the optimal one. The authors claimed that after comparing the three models based on their resulted accuracies, they found that the BCRSVM model outperformed the other two. From the initial set of 193 available variables in their dataset, only 14 features were selected based on their clinical knowledge. These data refer to clinical, epidemiological and pathological variables of 733 patients considered out of 1.541. In the final stage of the feature selection, Kaplan–Meier analysis and Cox regression were applied which resulted in 7 variables as most informative. These features were then entered as input to the SVM and ANN classifiers as well as to the Cox regression statistical model. In order to evaluate the performance of the models, the authors employed the hold-out method, which splits the data sample into two sub-sets, namely training and testing set. As in most studies in the literature, accuracy, sensitivity and specificity were calculated for a reliable estimation of the models. Based on these metrics, the authors claimed that BCRSVM outperformed the ANN and Cox regression models with accuracy 84.6%, 81.4% and 72.6%, respectively. Comparison among the performance of other previously established recurrence prediction models revealed that BCRSVM has superior performance. It should be noted that this study estimated also the importance of prognostic factors by means of normalized mutual information index (NMI) [61]. Based on these calculations for each of the three predictive models, they suggest that the most significant factor regarding the prediction of breast cancer recurrence was the local invasion of tumor. However, if someone reviews this work would certainly mention some major limitations. As the authors noted, the exclusion of a large number of patients (n = 808) due to the lack of clinical data in the research registry, influenced the performance of their models. Furthermore, the fact that the authors used only their clinical knowledge to select 14 out of 193 variables may have resulted in significant bias, thus giving no robust results. Apart from this limitation, the authors could also improve the performance of their proposed model, namely BCRSVM, by validating it with external datasets from other sources. Among the initial list of publications resulted from our literature survey, we noticed a growing trend the last years regarding the prediction of cancer disease by means of SSL learning. So, we believed it would be of interest to present the most recent study that makes use of this type of ML techniques for the analysis of breast cancer recurrence [48]. The proposed algorithm is based on the use of SSL for the construction of a graph model while it integrates gene expression data with gene network information in order to predict cancer recurrence. Based on biological knowledge, the authors selected gene pairs that indicate strong biological interactions. The sub-gene network identified by the proposed method is constituted of the BRCA1, CCND1, STAT1 and CCNB1 genes. Their methodology is divided in three sections including: (i) the determination of gene pairs for building the graph model with only labeled samples, (ii) the development of sample graphs based on informative genes and (iii) the regularization of the graph resulting in finding the labels of unlabeled samples. The dataset used through this study consists of gene expression profiles found in the GEO repository [62] as well as of PPIs derived from the I2D database[63]. Specifically, five gene expression datasets were downloaded from GEO including 125, 145, 181, 249 and 111 labeled samples. These samples were classified into three groups: (i) recurrence, (ii) non-recurrence and (iii) unlabeled samples and referred to cancer types like breast and colon cancer. Additionally, they downloaded from the I2D database a sample of human PPIs composed of 194.988 known, experimental and predicted interactions. After removing the duplicated PPIs and the interactions that do not contain proteins mapped to a gene they resulted in an amount of 108,544 interactions. Based on the results of this study, the authors showed that the gene networks derived from the SSL learning method include many important genes related to cancer recurrence. They also claimed that their approach outperforms other existing methods in the case of breast cancer recurrence prediction. The estimated performance of the proposed method compared to other known methods that make use of PPIs for the identification of informative genes showed an accuracy of 80.7% and 76.7% in the breast and colon cancer samples, respectively. Ten-fold cross validation was used for estimating the experimental results. Although this type of ML methods differs considerably from these of supervised and unsupervised learning on the algorithms that they employ, it is clear that it provides more advantages relevant to the collection of datasets and their sizes. Unlabeled data are cheap and can be easier extracted. On the contrary, labeled samples may require experts and special devices in order to be collected. This study reveals that SSL can be an alternative to supervised approaches which usually suffers from small labeled samples.

Prediction of Cancer Survival

In [26] a predictive model is developed for the evaluation of survival in women that have been diagnosed with breast cancer, while they addressed the importance of robustness under the model's parameter variation. They compared three classification models namely SVM, ANN and SSL based on the SEER cancer database [64]. The dataset is composed of 162,500 records with 16 key features. A class variable was also considered, namely survivability, referring to patients that had not survived and those that had survived. Among the most informative features are (i) the tumor size, (ii) the number of nodes and (iii) the age at diagnosis. By comparing the best performance of each of the three models they found that the calculated accuracy for ANN, SVM and SSL was 65%, 51% and 71% respectively. Five-fold cross validation was used for evaluating the performance of the predictive models. Concerning those findings the authors proposed the SSL model as a good candidate for survival analysis by the clinical experts. We should note that no preprocessing steps were mentioned by the authors regarding the collection of the most informative features. They proceeded with the entire SEER datasets and the box-whisper-plot was used for estimating the performance variation across 25 combinations of model parameters. A small box area of a specific model indicates more robustness and stability under parameter combination. The small boxes of the SSL model revealed its better accuracy than the other models. A relevant study was published the next year which attempts to assess the survival prediction of non-small cell lung cancer (NSCLC) patients through the use of ANNs [50]. Their dataset consists of NSCLC patients' gene expression raw data and clinical data obtained from the NCI caArray database [65]. After the preprocessing steps in their approach, the authors selected the most informative survival-associated gene signatures; LCK and ERBB2 genes, which were then used for training the ANN network. Four clinical variables, namely sex, age, T_stage and N_stage were also considered as input variables in the ANN model. They also performed several types of ANN architectures in order to find the optimal one for the prediction of cancer survival. An overall accuracy of 83% was provided regarding the predictive performance of the classification scheme. Furthermore, their results revealed that all patients were classified in different groups regarding their treatment protocol while 50% of them had not survived. The evaluation of the model outcomes was done based on the Kaplan–Meier survival analysis. They estimated the survival of patients for the training set, the test set and the validation set with p-value < 0.00001, while they showed that the patients in the high-risk group exhibited a lower median overall survival in comparison to low-risk patients. Compared to other studies in the literature relevant to NSCLC survival prediction, this work provided more stable results. However, existing limitations of the current article are related to the fact that the impact of other variables related to death (such as blood clots) is not considered, which may have led to misclassification results. Furthermore, the authors claim that their model could not be applied to other cancer types except NSCLC. This assumption is considered as a major limitation in studies that the predictive models may not generalize to different cancer types.

DISCUSSION

In the present review, the most recent works relevant to cancer prediction/ prognosis by means of ML techniques are presented. After a brief description of the ML branch and the concepts of the data preprocessing methods, the feature selection techniques and the classification algorithms being used, we outlined three specific case studies regarding the prediction of cancer susceptibility, cancer recurrence and cancer survival based on popular ML tools. Obviously, there is a large amount of ML studies published in the last decade that provide accurate results concerning the specific predictive cancer outcomes. However, the identification of potential drawbacks including the experimental design, the collection of appropriate data samples and the validation of the classified results, is critical for the extraction of clinical decisions.

Moreover, it should be mentioned that in spite of the claims that these ML classification techniques can result in adequate and effective decision making, very few have actually penetrated the clinical practice. Recent advances in omics technologies paved the way to further improve our understanding of a variety of diseases; however more accurate validation results are needed before gene expression signatures can be useful in the clinics.

A growing trend was noted in the studies published the last 2 years that applied semi-supervised ML techniques for modeling cancer survival. This type of algorithms employs labeled and unlabeled data for their predictions while it has been proven that they improved the estimated performance compared to existing supervised techniques [26]. SSL can be though as a great alternative to the other two types of ML methods (i.e. supervised learning and unsupervised learning) that use, in general, only a few labeled samples.

One of the most common limitations noted in the studies surveyed in this review is the small amount of data samples. A basic requirement when using classification schemes for modeling a disease is the size of the training datasets that needs to be sufficiently large. A relatively large dataset allows the sufficient partitioning into training and testing sets, thus leading to reasonable validation of the estimators. A small sized training sample, compared to data dimensionality, can result in misclassifications while the estimators may produce unstable and biased models. It is obvious that a richer set of patients used for their survival prediction can enhance the generalizability of the predictive model.

Except the data size, the dataset quality as well as the careful feature selection schemes are of great importance for effective ML and subsequently for accurate cancer predictions. Choosing the most informative feature subset for training a model, by means of feature selection methods, could result in robust models. Additionally, feature sets that consist of histological or pathological assessments are characterized by reproducible values. Due to the lack of static entities when dealing with clinical variables it is important for a ML technique to be adjusted to different feature sets over time.

It should be noted that almost all of the works presented here, performed validation tests for estimating the performance of their learning algorithms. They employed well-known evaluation techniques that split the initial datasets into subsets. As mentioned above, in order to obtain accurate results for their predictive models, the authors should select large and independent features that could result in better validation. Internal and external validation was performed in these studies that would enable the extraction of more accurate and reliable predictions while it would minimize any bias [47].

A key point to several studies, regarding their promising results, was the fact that several ML techniques were employed as an aim to find the most optimal one [34]. Apart from this, the combination of multiple data types that would be fed as input to the models is also a trend. Looking back to the previous decade, only molecular and clinical information was exploited for making predictions of cancer outcomes. With the rapid development of HTTs, including genomic, proteomic and imaging technologies, new types of input parameters have been collected. We found that almost all the predictions was made by integrating either genomic, clinical, histological, imaging, demographic, epidemiological data and proteomic data or different combinations of these types [24]; [26]; [48]; [50] ; [53].

Additionally, there has been considerable activity regarding the integration of different types of data in the field of breast cancer [66] ; [67]. In the DREAM project [68], several attempts to combine clinical treatment scores with signatures based on immunohistochemistry [69]as well as expression-based signatures such as PAM50 [70] and Oncotype DX [71] reveal the extensive work done for improving treatment based on the incorporation of different features.

Among the most common applied ML algorithms relevant to the prediction outcomes of cancer patients, we found that SVM and ANN classifiers were widely used. As mentioned to our introductory section, ANNs have been used extensively for nearly 30 years [30]. In addition,

SVMs constitute a more recent approach in the cancer prediction/ prognosis and have been used widely due to its accurate predictive performance. However, the choice of the most appropriate algorithm depends on many parameters including the types of data collected, the size of the data samples, the time limitations as well as the type of prediction outcomes.

Concerning the future of cancer modeling new methods should be studied for overcoming the limitations discussed above. A better statistical analysis of the heterogeneous datasets used would provide more accurate results and would give reasoning to disease outcomes. Further research is required based on the construction of more public databases that would collect valid cancer dataset of all patients that have been diagnosed with the disease. Their exploitation by the researchers would facilitate their modeling studies resulting in more valid results and integrated clinical decision making.

CONCLUSIONS

In this review, we discussed the concepts of ML while we outlined their application in cancer prediction/prognosis. Most of the studies that have been proposed the last years and focus on the development of predictive models using supervised ML methods and classification algorithms aiming to predict valid disease outcomes. Based on the analysis of their results, it is evident that the integration of multidimensional heterogeneous data, combined with the application of different techniques for feature selection and classification can provide promising tools for inference in the cancer domain.

ACKNOWLEDGEMENTS

This work was part funded by the project NEOMARK (FP7-ICT 224483, ICT Enabled Prediction of Cancer Reoccurrence).

REFERENCES

1. Hanahan D, Weinberg RA. Hallmarks of cancer: the next generation. Cell 2011;144: 646–74
2. M.-Y.C. Polley, B. Freidlin, E.L. Korn, B.A. Conley, J.S. Abrams, L.M. McShane Statistical and practical considerations for clinical evaluation of predictive biomarkers J Natl Cancer Inst, 105 (2013), pp. 1677–1683
3. J.A. Cruz, D.S. Wishart Applications of machine learning in cancer prediction and prognosis Cancer Informat, 2 (2006), p. 59
4. O. Fortunato, M. Boeri, C. Verri, D. Conte, M. Mensah, P. Suatoni, et al. Assessment of circulating microRNAs in plasma of lung cancer patients Molecules, 19 (2014), pp. 3038–3054
5. H.M. Heneghan, N. Miller, M.J. Kerin MiRNAs as biomarkers and therapeutic targets in cancer Curr Opin Pharmacol, 10 (2010), pp. 543–550
6. D. Madhavan, K. Cuk, B. Burwinkel, R. Yang Cancer diagnosis and prognosis decoded by blood-based circulating microRNA signatures Front Genet, 4 (2013)
7. K. Zen, C.Y. Zhang Circulating microRNAs: a novel class of biomarkers to diagnose and monitor human cancers Med Res Rev, 32 (2012), pp. 326–348
8. S. Koscielny Why most gene expression signatures of tumors have not been useful in the clinic Sci Transl Med, 2 (2010) [14 ps12-14 ps12]
9. S. Michiels, S. Koscielny, C. Hill Prediction of cancer outcome with microarrays: a multiple random validation strategy Lancet, 365 (2005), pp. 488–492
10. C.M. Bishop Pattern recognition and machine learning Springer, New York (2006)
11. T.M. Mitchell The discipline of machine learning: Carnegie Mellon University Carnegie Mellon University, School of Computer Science, Machine Learning Department (2006)
12. I.H. Witten, E. Frank Data mining: practical machine learning tools and techniques Morgan Kaufmann (2005)
13. A Niknejad, D. Petrovic Introduction to computational intelligence

techniques and areas of their applications in medicine Med Appl Artif Intell, 51 (2013)

14. T. Pang-Ning, M. Steinbach, V. Kumar Introduction to data mining (2006)
15. Y. Drier, E. Domany Do two machine-learning based prognostic signatures for breast cancer capture the same biological processes? PLoS One, 6 (2011), p. e17795
16. A Dupuy, R.M. Simon Critical review of published microarray studies for cancer outcome and guidelines on statistical analysis and reporting J Natl Cancer Inst, 99 (2007), pp. 147–157
17. L. Ein-Dor, I. Kela, G. Getz, D. Givol, E. Domany Outcome signature genes in breast cancer: is there a unique set? Bioinformatics, 21 (2005), pp. 171–178
18. L. Ein-Dor, O. Zuk, E. Domany Thousands of samples are needed to generate a robust gene list for predicting outcome in cancer Proc Natl Acad Sci, 103 (2006), pp. 5923–5928
19. T. Ayer, O. Alagoz, J. Chhatwal, J.W. Shavlik, C.E. Kahn, E.S. Burnside Breast cancer risk estimation with artificial neural networks revisited Cancer, 116 (2010), pp. 3310–3321
20. J.C. Platt, N. Cristianini, J. Shawe-Taylor Large margin DAGs for multiclass classification (1999), pp. 547–553
21. S. Adams Is Coursera the beginning of the end for traditional higher education? Higher Education (2012)
22. D. Cicchetti Neural networks and diagnosis in the clinical laboratory: state of the art Clin Chem, 38 (1992), pp. 9–10
23. A.J. Cochran Prediction of outcome for patients with cutaneous melanoma Pigment Cell Res, 10 (1997), pp. 162–167
24. K.P. Exarchos, Y. Goletsis, D.I. Fotiadis Multiparametric decision support system for the prediction of oral cancer reoccurrence IEEE Trans Inf Technol Biomed, 16 (2012), pp. 1127–1134
25. Kononenko Machine learning for medical diagnosis: history, state of the art and perspective Artif Intell Med, 23 (2001), pp. 89–109
26. K. Park, A. Ali, D. Kim, Y. An, M. Kim, H. Shin Robust predictive model for evaluating breast cancer survivability Engl Appl Artif Intell, 26 (2013), pp. 2194–2205

27. Y. Sun, S. Goodison, J. Li, L. Liu, W. Farmerie Improved breast cancer prognosis through the combination of clinical and genetic markers Bioinformatics, 23 (2007), pp. 30–37
28. L. Bottaci, P.J. Drew, J.E. Hartley, M.B. Hadfield, R. Farouk, P.WR. Lee, et al. Artificial neural networks applied to outcome prediction for colorectal cancer patients in separate institutions Lancet, 350 (1997), pp. 469–472
29. P.S. Maclin, J. Dempsey, J. Brooks, J. Rand Using neural networks to diagnose cancer J Med Syst, 15 (1991), pp. 11–19
30. R.J. Simes Treatment selection for cancer patients: application of statistical decision theory to the treatment of advanced ovarian cancer J Chronic Dis, 38 (1985), pp. 171–186
31. M.F. Akay Support vector machines combined with feature selection for breast cancer diagnosis Expert Syst Appl, 36 (2009), pp. 3240–3247
32. S.-W. Chang, S. Abdul-Kareem, A.F. Merican, R.B. Zain Oral cancer prognosis based on clinicopathologic and genomic markers using a hybrid of feature selection and machine learning methods BMC Bioinforma, 14 (2013), p. 170
33. L.-Y. Chuang, K.-C. Wu, H.-W. Chang, C.-H. Yang Support vector machine-based prediction for oral cancer using four snps in DNA repair genes (2011), pp. 16–18
34. A.T. Eshlaghy, A. Poorebrahimi, M. Ebrahimi, A.R. Razavi, L.G. Ahmad Using three machine learning techniques for predicting breast cancer recurrence J Health Med Inform, 4 (2013), p. 124
35. K.P. Exarchos, Y. Goletsis, D.I. Fotiadis A multiscale and multiparametric approach for modeling the progression of oral cancer BMC Med Inform Decis Mak, 12 (2012), p. 136
36. J. Kim, H. Shin Breast cancer survivability prediction using labeled, unlabeled, and pseudo-labeled patient data J Am Med Inform Assoc, 20 (2013), pp. 613–618
37. L.P. Fielding, C.M. Fenoglio-Preiser, L.S. Freedman The future of prognostic factors in outcome prediction for patients with cancer Cancer, 70 (1992), pp. 2367–2377
38. P.B. Bach, M.W. Kattan, M.D. Thornquist, M.G. Kris, R.C. Tate,

M.J. Barnett, et al. Variations in lung cancer risk among smokers J Natl Cancer Inst, 95 (2003), pp. 470–478

39. S.M. Domchek, A. Eisen, K. Calzone, J. Stopfer, A. Blackwood, B.L. Weber Application of breast cancer risk prediction models in clinical practice J Clin Oncol, 21 (2003), pp. 593–601
40. F. Gascon, M. Valle, R. Martos, M. Zafra, R. Morales, M.A. Castano Childhood obesity and hormonal abnormalities associated with cancer risk Eur J Cancer Prev, 13 (2004), pp. 193–197
41. X. Ren, Y. Wang, L. Chen, X.-S. Zhang, Q. Jin ellipsoidFN: a tool for identifying a heterogeneous set of cancer biomarkers based on gene expressions Nucleic Acids Res, 41 (2013), p. e53
42. X. Ren, Y. Wang, X.-S. Zhang, Q. Jin iPcc: a novel feature extraction method for accurate disease class discovery and prediction Nucleic Acids Res (2013), p. gkt343
43. Y. Wang, Q.-F. Wu, C. Chen, L.-Y. Wu, X.-Z. Yan, S.-G. Yu, et al. Revealing metabolite biomarkers for acupuncture treatment by linear programming based feature selection BMC Syst Biol, 6 (2012), p. S15
44. M. Waddell, D. Page, J. Shaughnessy Jr. Predicting cancer susceptibility from single-nucleotide polymorphism data: a case study in multiple myeloma ACM (2005), pp. 21–28
45. J. Listgarten, S. Damaraju, B. Poulin, L. Cook, J. Dufour, A. Driga, et al. Predictive models for breast cancer susceptibility from multiple single nucleotide polymorphisms Clin Cancer Res, 10 (2004), pp. 2725–2737
46. A Stojadinovic, A. Nissan, J. Eberhardt, T.C. Chua, J.O.W Pelz, J. Esquivel Development of a Bayesian belief network model for personalized prognostic risk assessment in colon carcinomatosis Am Surg, 77 (2011), pp. 221–230
47. W. Kim, K.S. Kim, J.E. Lee, D.-Y. Noh, S.-W. Kim, Y.S. Jung, et al. Development of novel breast cancer recurrence prediction model using support vector machine J Breast Cancer, 15 (2012), pp. 230–238
48. C. Park, J. Ahn, H. Kim, S. Park Integrative gene network construction to analyze cancer recurrence using semi-supervised learning PLoS One, 9 (2014), p. e86309

49. C.-J. Tseng, C.-J. Lu, C.-C. Chang, G.-D. Chen Application of machine learning to predict the recurrence-proneness for cervical cancer Neural Comput & Applic, 24 (2014), pp. 1311–1316
50. Y.-C. Chen, W.-C. Ke, H.-W. Chiu Risk classification of cancer survival using ANN with gene expression data from multiple laboratories Comput Biol Med, 48 (2014), pp. 1–7
51. X. Xu, Y. Zhang, L. Zou, M. Wang, A. Li A gene signature for breast cancer prognosis using support vector machine IEEE (2012), pp. 928–931
52. O. Gevaert, F. De Smet, D. Timmerman, Y. Moreau, B. De Moor Predicting the prognosis of breast cancer by integrating clinical and microarray data with Bayesian networks Bioinformatics, 22 (2006), pp. e184–e190
53. P. Rosado, P. Lequerica-Fernández, L. Villallaín, I. Peña, F. Sanchez-Lasheras, J.C. de Vicente Survival model in oral squamous cell carcinoma based on clinicopathological parameters, molecular markers and support vector machines Expert Syst Appl, 40 (2013), pp. 4770–4776
54. D. Delen, G. Walker, A. Kadam Predicting breast cancer survivability: a comparison of three data mining methods Artif Intell Med, 34 (2005), pp. 113–127
55. R.J. Urbanowicz, A.S. Andrew, M.R. Karagas, J.H. Moore Role of genetic heterogeneity and epistasis in bladder cancer susceptibility and outcome: a learning classifier system approach J Am Med Inform Assoc, 20 (2013), pp. 603–612
56. A Bochare, A. Gangopadhyay, Y. Yesha, A. Joshi, Y. Yesha, M. Brady, et al. Integrating domain knowledge in supervised machine learning to assess the risk of breast cancer Int J Med Eng Inform, 6 (2014), pp. 87–99
57. S. Gilmore, R. Hofmann-Wellenhof, H.P. Soyer A support vector machine for decision support in melanoma recognition Exp Dermatol, 19 (2010), pp. 830–835
58. N. Mac Parthaláin, R. Zwiggelaar Machine learning techniques and mammographic risk assessment. Digital mammography Springer (2010) pp. 664–672

59. M.A. Hall Feature selection for discrete and numeric class machine learning (1999)

60. R. Kohavi, G.H. John Wrappers for feature subset selection Artif Intell, 97 (1997), pp. 273–324

61. P.A. Estévez, M. Tesmer, C.A. Perez, J.M. Zurada Normalized mutual information feature selection IEEE Trans Neural Netw, 20 (2009), pp. 189–201

62. T. Barrett, D.B. Troup, S.E. Wilhite, P. Ledoux, D. Rudnev, C. Evangelista, et al. NCBI GEO: mining tens of millions of expression profiles — database and tools update Nucleic Acids Res, 35 (2007), pp. D760–D765

63. Y. Niu, D. Otasek, I. Jurisica Evaluation of linguistic features useful in extraction of interactions from PubMed; application to annotating known, high-throughput and predicted interactions in I2D Bioinformatics, 26 (2010), pp. 111–119

64. N. Howlader, A. Noone, M. Krapcho, J. Garshell, N. Neyman, S. Aletkruse SEER Cancer Statistics Review, 1975–2010, [Online] National Cancer Institute National Cancer Institute, Bethesda, MD (2013) [Online]

65. X. Bian, J. Klemm, A. Basu, J. Hadfield, R. Srinivasa, T. Parnell, et al. Data submission and curation for caArray, a standard based microarray data repository system (2009)

66. A Papadopoulos, D.I. Fotiadis, L. Costaridou Improvement of microcalcification cluster detection in mammography utilizing image enhancement techniques Comput Biol Med, 38 (2008), pp. 1045–1055

67. A Papadopoulos, D.I. Fotiadis, A. Likas Characterization of clustered microcalcifications in digitized mammograms using neural networks and support vector machines Artif Intell Med, 34 (2005), pp. 141–150

68. E. Bilal, J. Dutkowski, J. Guinney, I.S. Jang, B.A. Logsdon, G. Pandey, et al. Improving breast cancer survival analysis through competition-based multidimensional modeling PLoS Comput Biol, 9 (2013), p. e1003047

69. J. Cuzick, M. Dowsett, S. Pineda, C. Wale, J. Salter, E. Quinn, et al. Prognostic value of a combined estrogen receptor, progesterone receptor, Ki-67, and human epidermal growth factor receptor 2 immunohistochemical score and comparison with the Genomic Health recurrence score in early breast cancer J Clin Oncol, 29 (2011), pp. 4273–4278
70. J.S. Parker, M. Mullins, M.C. Cheang, S. Leung, D. Voduc, T. Vickery, et al. Supervised risk predictor of breast cancer based on intrinsic subtypes J Clin Oncol, 27 (2009), pp. 1160–1167
71. S. Paik, S. Shak, G. Tang, C. Kim, J. Baker, M. Cronin, et al. A multigene assay to predict recurrence of tamoxifen-treated, node-negative breast cancer N Engl J Med, 351 (2004), pp. 2817–2826

5 MACHINE LEARNING AND DATA MINING METHODS IN DIABETES RESEARCH

Ioannis Kavakiotis [a,b], Olga Tsave [c], Athanasios Salifoglou [c], Nicos Maglaveras [b,d],Ioannis Vlahavas [a], Ioanna Chouvarda [b,d]

[a] Department of Informatics, Aristotle University of Thessaloniki, Thessaloniki 54124, Greece

[b] Institute of Applied Biosciences, CERTH, Thessaloniki, Greece

[c] Laboratory of Inorganic Chemistry, Department of Chemical Engineering, Aristotle University of Thessaloniki, Thessaloniki 54124, Greece

[d] Lab of Computing and Medical Informatics, Medical School, Aristotle University of Thessaloniki, Thessaloniki 54124, Greece

ABSTRACT

The remarkable advances in biotechnology and health sciences have led to a significant production of data, such as high throughput genetic data and clinical information, generated from large Electronic Health Records (EHRs). To this end, application of machine learning and data mining methods in biosciences is presently, more than ever before, vital and indispensable in efforts to transform intelligently all available information into valuable knowledge. Diabetes mellitus (DM) is defined as a group of metabolic disorders exerting significant pressure on human health worldwide. Extensive research in all aspects of diabetes (diagnosis, etiopathophysiology, therapy, etc.) has led to the generation of huge amounts of data. The aim of the present study is to conduct a systematic review of the applications of machine learning, data mining techniques and tools in the field of diabetes research with respect to a) Prediction and Diagnosis, b) Diabetic Complications, c) Genetic Background and Environment, and e) Health Care and Management with the first category appearing to be the most popular. A wide range of machine learning algorithms were employed. In general, 85% of those used were characterized by supervised learning approaches and 15% by unsupervised ones, and more specifically, association rules. Support vector machines (SVM) arise as the most successful and widely used algorithm. Concerning the type of data, clinical datasets were mainly used. The title applications in the selected articles project the usefulness of extracting valuable knowledge leading to new hypotheses targeting deeper understanding and further investigation in DM.

Keywords: Machine learning, Data mining, Diabetes mellitus, Diabetic complications, Disease prediction models, Biomarker(s) identification

INTRODUCTION

Significant advances in biotechnology and more specifically high-throughput sequencing result incessantly in an easy and inexpensive data production, thereby ushering the science of applied biology into the area of big data [1,2].

To date, besides high performance sequencing methods, there is a plethora of digital machines and sensors from various research fields generating data, including super-resolution digital microscopy, mass spectrometry, Magnetic Resonance Imagery (MRI), etc. Although these technologies produce a wealth of data, they do not provide any kind of analysis, interpretation or extraction of knowledge. To this end, the area of Biological Data Mining or otherwise Knowledge Discovery in

Biological Data, is more than ever necessary and important. The primary objective is to delve into the rapidly accruing body of biological data and set the basis potentiating answers to fundamental questions in biology and medicine.

The power and effectiveness of these approaches are derived from the ability of commensurate methods to extract patterns and create models from data. The aforementioned fact is particularly significant in the big data era, especially when the dataset can reach terabytes or petabytes of data. Consequently, the abundance of data has strengthened considerably data-oriented research in biology. In such a hybrid field, one of the most important research applications is prognosis and diagnosis related to human-threatening and/or life quality reducing diseases. One such disease is diabetes mellitus (DM).

Applying machine learning and data mining methods in DM research is a key approach to utilizing large volumes of available diabetes-related data for extracting knowledge. The severe social impact of the specific disease renders DM one of the main priorities in medical science research, which inevitably generates huge amounts of data. Undoubtedly, therefore, machine learning and data mining approaches in DM are of great concern when it comes to diagnosis, management and other related clinical administration aspects. Hence, in the framework of this study, efforts were made to review the current literature on machine learning and data mining approaches in diabetes research.

The review is organized as follows: Section 2 provides the necessary background knowledge on machine learning (ML) and knowledge discovery in databases (KDD). Section 3 presents a concise presentation of the DM disease. Section 4 provides the methodological approach adopted, and Section 5, divided in five subsections, presents publications reviewed in the study. Section 6 presents a discussion, with Section 7 providing conclusions.

MACHINE LEARNING AND KNOWLEDGE DISCOVERY

Machine learning is the scientific field dealing with the ways in which machines learn from experience. For many scientists, the term "machine learning" is identical to the term "artificial intelligence", given that the possibility of learning is the main characteristic of an entity called intelligent in the broadest sense of the word. The purpose of machine learning is the construction of computer systems that can adapt and learn from their experience [3]. A more detailed and formal definition of machine learning is given by Mitchel [4]: *A computer program is said to learn from experience E with respect to some class of tasks T and performance measure P, if its performance at tasks in T, as measured by P, improves with experience E.*

Knowledge discovery in databases (KDD) is a field encompassing theories, methods and techniques, trying to make sense of data and extract useful knowledge from them. It is considered to be a multistep process (selection, preprocess, transformation, data mining, interpretation-evaluation) depicted in Fig. 1[5]. The most important step in the entire KDD process is data mining, exemplifying the application of machine learning algorithms in analyzing data. A complete definition of KDD is given by Fayyad et al. [5]: *KDD is the nontrivial process identifying valid, novel, potentially useful, and ultimately understandable patterns in data.*

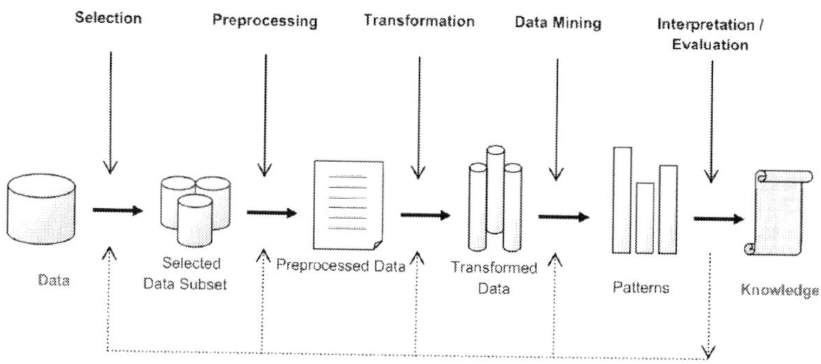

Figure 1: The basic steps of the KDD process.

Categories of Machine Learning Tasks

Machine learning tasks are typically classified into three broad categories [6]. These are: a) supervised learning, in which the system infers a function from labeled training data, b) unsupervised learning, in which the learning system tries to infer the structure of unlabeled data, and c) reinforcement learning, in which the system interacts with a dynamic environment.

Supervised Learning

In supervised learning, the system must "learn" inductively a function called target function, which is an expression of a model describing the data. The objective function is used to predict the value of a variable, called dependent variable or output variable, from a set of variables, called independent variables or input variables or characteristics or features. The set of possible input values of the function, i.e. its domain, are called instances. Each case is described by a set of characteristics (attributes or features). A subset of all cases, for which the output variable value is known, is called training data or examples. In order to infer the best target function, the learning system, given a training set, takes into consideration alternative functions, called hypothesis and denoted by *h*. In supervised learning, there are two kinds of learning tasks: classification and regression. Classification models try to predict distinct classes, such as e.g. blood groups, while regression models predict numerical values. Some of the most common techniques are Decision Trees (DT), Rule Learning, and Instance Based Learning (IBL), such as *k*-Nearest Neighbors (*k*-NN), Genetic Algorithms (GA), Artificial Neural Networks (ANN), and Support Vector Machines (SVM).

Unsupervised Learning

In unsupervised learning, the system tries to discover the hidden structure of data or associations between variables. In that case, training data consists of instances without any corresponding labels.

Association Rule Learning

Association Rule Mining appeared much later than machine learning and is subject to greater influence from the research area of databases. It was proposed in the early 1990s by Rakesh Agrawal [7] as a market basket analysis, in which the aim was to find correlations in the objects of a database. Based on the shopping cart example, association rules are of the form $\{X_1, \ldots, X_n\} \rightarrow Y$, which means that if you find all of $X_1, \ldots, X_n$ in a cart it is *possible* to find Y. The most well-known association rule discovery algorithm is *Apriori*, proposed in 1994 by Rakesh Agrawal [8].

Although, association rule mining was first introduced as a market basket analysis tool, it has since become one of the most valuable tools for performing unsupervised exploratory data analysis over a wide range of research and commercial areas, including biology and bioinformatics. Some of the most well-known applications in biology and bioinformatics include biological sequence analysis, analysis of gene expression data and others. A thorough review of discovering frequent patterns and association rules from biological data, including algorithms and applications, can be found in [9].

Clustering

Clusters are informative patterns occurring through clustering, i.e. the separation of a whole dataset into groups of data, so that instances belonging to the same group are as similar as possible and instances belonging to different groups differ as much as possible [10].

Reinforcement Learning

The term *Reinforcement Learning* is a general term given to a family of techniques, in which the system attempts to learn through direct interaction with the environment so as to maximize some notion of cumulative reward [11]. It is important to mention that the system has no prior knowledge about the behavior of the environment and the only way to find out is through trial and failure (trial and error). Reinforcement learning is mainly applied to autonomous systems, due to its independence in relation to its environment.

Feature Selection

Feature selection is one of the most important processes of the KDD's data transformation step. It is defined as the process of selecting a subset of features from the feature space, which is more relevant to and informative for the construction of a model. The advantages of feature selection are many and relate to different aspects of data analysis, such as better visualization and understanding of data, reduction of computational time and duration of analysis, and better prediction accuracy [12,13].

There are two main different approaches in the feature selection process. The first one is to make an independent assessment, based on general characteristics of data. Methods belonging to this approach are called filter methods, because the feature set is filtered out before model construction. The second approach is to use a machine learning algorithm to evaluate different subsets of features and finally select the one with the best performance on classification accuracy. The latter algorithm will be used in the end to build a predictive model. Methods in this category are called wrapper methods, because the arising algorithm wraps the whole feature selection process.

DIABETES MELLITUS

Diabetes Mellitus (DM) is defined as a group of metabolic disorders mainly caused by abnormal insulin secretion and/or action [14]. Insulin deficiency results in elevated blood glucose levels (hyperglycemia) and impaired metabolism of carbohydrates, fat and proteins. DM is one of the most common endocrine disorders, affecting more than 200 million people worldwide. The onset of diabetes is estimated to rise dramatically in the upcoming years. DM can be divided into several distinct types. However, there are two major clinical types, type 1 diabetes (T1D) and type 2 diabetes (T2D), according to the etiopathology of the disorder. T2D appears to be the most common form of diabetes (90% of all diabetic patients), mainly characterized by insulin resistance. The main causes of T2D include lifestyle, physical activity, dietary habits and heredity, whereas T1D is thought to be due to autoimmunological destruction of the Langerhans islets hosting pancreatic-β cells. T1D affects almost 10% of all diabetic patients worldwide, with 10% of them ultimately developing idiopathic diabetes. Other forms of DM, classified on the basis of insulin secretion profile and/or onset, include Gestational Diabetes, endocrinopathies, MODY (Maturity Onset Diabetes of the Young), neonatal, mitochondrial, and pregnancy diabetes. The symptoms of DM include polyurea, polydipsia, and significant weight loss among others. Diagnosis depends on blood glucose levels (fasting plasma glucose = 7.0 mmol/L) [15].

DM progression is strongly linked to several complications, mainly due to chronic hyperglycemia. It is well-known that DM covers a wide range of heterogeneous pathophysiological conditions. The most common complications are divided into micro- and macro-vascular disorders, including diabetic nephropathy, retinopathy, neuropathy, diabetic coma and cardiovascular disease. Due to high DM mortality and morbidity as well as related disorders, prevention and treatment attracts broad and significant interest. Insulin administration is the main treatment for T1D, although insulin is also provided in certain cases of T2D patients, when hyperglycemia cannot be controlled through diet, weight loss, exercise and oral medication. Current medication targets primarily a) saving one's life and alleviating the disease symptoms, and b) prevention of long term diabetic complications and/or elimination of several risk factors, thereby increasing longevity. The most common anti-diabetic agents include sulfonylurea, metformin, alpha-glucosidase inhibitor, peptide analogs, non-sulfonylurea secretagogues, etc. [16]. The majority of the present anti-diabetic agents, however, exhibit numerous side-effects. In addition, insulin therapy is related to weight gain and hypoglycemic events. Hence, anti-diabetic drug design and discovery is of great concern and concurrently a research challenge [17–20].

Although extensive research in DM has provided significant knowledge, over the past decades, on the a) etiopathology (genetic or environmental factors and cellular mechanisms), b) treatment, and c) screening and management of the disease, there is still much to be discovered, unraveled, clarified and delineated. Through such processes, diagnosis, prognostic evaluation of appropriate treatment and clinical administration could gain significant ground toward medical handling of the disease. In such an effort, reliance on a large and fast increasing body of research and clinical data serves to establish a significant basis for safe diagnosis and follow-up treatment. Thus, data mining and machine learning emerge as key processes, contributing decisively to the decision-making clinician. The aspiration, therefore, is to link data assessment to diagnosis and appropriate decision-making in drug administration.

METHODS

Extensive efforts were made to identify articles employing machine learning and data mining techniques on diabetes research. Two databases were searched (15 July 2016): the one extensively used in biomedical sciences, PubMed and the DBLP Computer Science Bibliography, containing more than 3.4 million journal articles, conference papers, and other publications on computer science (July 2016) [21]. The main reason behind the utilization of DBLP was that there are certain high impact international scientific journals in the computer science field that are not indexed by PubMed, although in some cases, the proposed published methods are applied on biomedical datasets.

As mentioned previously, there is a close relationship between the terms machine learning and data mining, with the latter being more generic. Thus, often, in scientific literature, machine learning methods are called data mining methods. To overcome that and be more accurate in finding all related articles, two searches were performed in PubMed, based on the following queries: a) “Machine Learning” AND “Diabetes” (QUERY_1), and b) “Data Mining” AND “Diabetes” (QUERY_2). Although PubMed launches searches on the title, abstracts and keywords of an article, DBLP conducts searches only on the title. In view of this fact, searches in DBLP were limited only to “Diabetes” query (QUERY_3), since machine learning and data mining are too broad terms to be found on a computer science article title.

Due to the vast amount of articles returned from the three queries (QUERY_1: 139, QUERY_2: 268, and QUERY_3: 880), our search was limited to articles published over the past five years (automatically in PubMed and manually in DBLP), thereby narrowing down significantly the retrieved collection (QUERY_1: 110, QUERY_2: 184, and QUERY_3: 248). It is important to mention that the huge collection of articles, retrieved through DBLP, was due to the fact that articles were not only limited to the machine learning and data mining fields but also covered the broader computer science field in general.

The next step was manual inspection of all recovered articles. The common purpose of this manual inspection for all three queries was to initially assess their relationship to Diabetes research. Moreover, for QUERY_2, manual inspection was performed to exclude articles that didn’t contain machine learning methods; for instance, articles with simple statistical analyses. Lastly, concerning QUERY_3, the purpose of manual inspection was twofold. Firstly, to find all articles related to machine learning and secondly to identify and merge overlaps among queries, i.e. articles already indexed by PubMed, which included the vast majority of them. Manual inspection narrowed down even more the collection (QUERY_1: 54, QUERY_2: 36, and QUERY_3: 13), thereby resulting in a final collection of 103 articles. These articles were classified into the following five categories: Biomarker Prediction and Diagnosis in DM; Diabetic Complications; Drugs and Therapies; Genetic Background and Environment, and Health Care Management. The entire article selection process is illustrated in a workflow (Fig. 2), with the number of publications per year being depicted in Fig. 3.

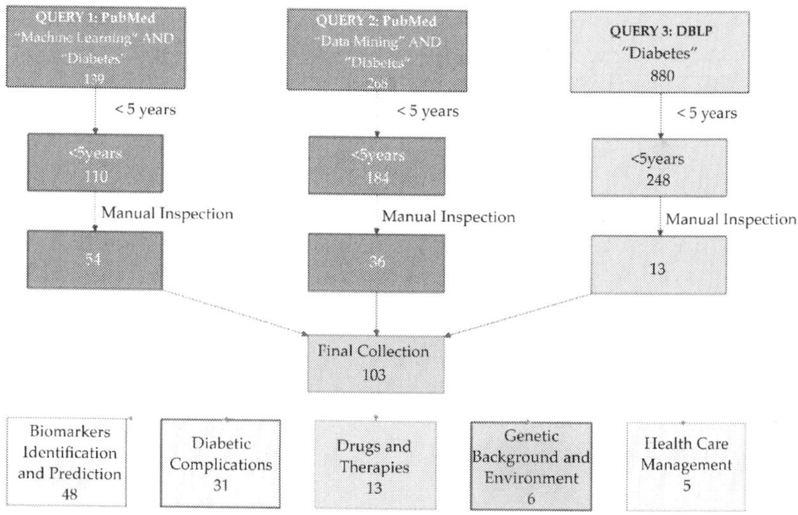

Figure 2: Literature selection and classification process.

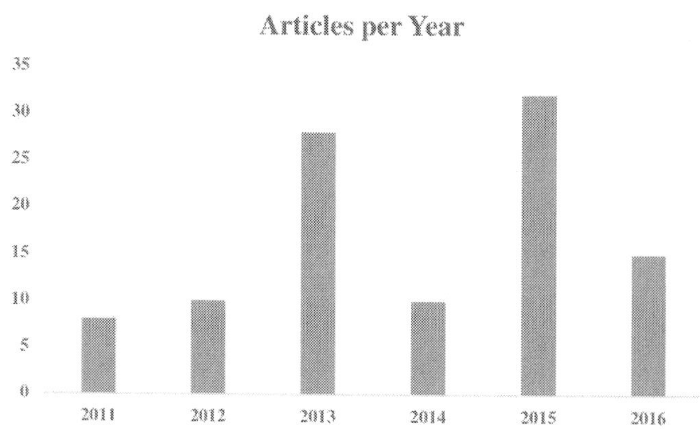

Figure 3: Articles per year in the collection employed.

Since the area of data mining and machine learning applied to Diabetes is very wide, it is hard to include every single research study. The selected methodology was employed in an effort to present only the latest research efforts in DM. In this term, the current collection consists of research work conducted the last five years. Moreover, in the present study, specific keywords were used such as "machine learning" and "data mining". However, there are several additional keywords that could possibly be used concerning specific algorithms, e.g. neural networks or specific tasks (i.e. predictive modeling), that belong to the field of machine learning and data mining without mentioning the current terms. In that sense, the results of this research are not exhaustive.

DM THROUGH MACHINE LEARNING AND DATA MINING

This section presents key papers of the study.

Biomarker Identification and Prediction of DM

A large number of factors are known to be important in the development and progression of DM. Obesity stands as a major risk factor, especially in T2D, given the strong causal relationship between that and the onset of DM [22]. DM diagnosis is carried out through several tests [α-glycate hemoglobin (A1C) test, random blood sugar test, fasting sugar test or oral glucose tolerance test]. There is evidence that in both T1D and T2D, early diagnosis and prediction of the onset of the disease are vital to the a) retardation of the progression of the disease, b) targeted selection of the medication, c) prolonging life expectancy, symptom alleviation, and d) onset of related complications.

Biomarkers (e.g. biological molecules) are measurable indicators of a certain condition representing health and disease states. Typically, biomarkers are a) measured in body fluids (blood, saliva or urine), b) encountered and thus determined independent of their etiopathogenic mechanistic pathway, and c) used to monitor clinical and subclinical disease burden and response to treatments. Biomarkers can be direct ending points of the disease itself or indirect indexes of other complications. Current technologies, such as metabolomics, proteomics, and genomics contribute to the development of a plethora of new biomarkers. In the case of DM, biomarkers may reflect the presence and severity of hyperglycemia or presence and severity of the related complications in diabetes [23].

The current section is divided into two main categories, which include cases where a) diagnostic and predictive markers are employed or new biomarkers are introduced, and b) disease prediction takes place, although this task is always performed to evaluate the predictive accuracy of the identified biomarkers.

Diagnostic and Predictive Markers

The first category deals with biomarker discovery, which is a task mainly performed through feature selection techniques [24–34]. Following a feature selection step, a classification algorithm is employed to assess the prediction accuracy of the selected features.

Firstly, established methods have been used in the biomarker evaluation issue. In [25], the authors used a clinical dataset comprised of 803 prediabetic females with 55 features and compared several common feature selection algorithms (both wrapper and filter methods) to predict DM. They concluded that the best overall performance had been achieved through wrapper methods. Moreover, among the filter methods used, symmetrical uncertainty achieved the best prediction accuracy. In another work, using established methods, Georga et al. [28] applied Random Forest (RF) [35] and RReliefF [36] to evaluate a number of features, with respect to their ability to predict the short-term subcutaneous glucose concentrations. In [31], authors combined gas chromatography–mass spectrometry (GC/MS) profiling with Random Forest, in an effort to explore relationships between 5′-AMP-activated protein kinase AMPK and DM. Jelinek et al. [24] investigated whether additional biomarkers could be used together with HbA1c to improve diagnostic accuracy in T2D, in case HbA1c levels are below or equal to the current cut-off of 6.5%. They concluded that both 8-hydroxy-2-deoxyguanosine (8-OhdG), an oxidative stress marker, and interleukin-6 (IL-6) improved classification accuracy.

Novel methods have also been proposed to deal with features in diabetic patient data. Improved electromagnetism-like mechanism (IEM) algorithm [26] was proposed for feature selection. It combines electromagnetism-like mechanism (EM) algorithm with the nearest neighbor classifier [37] and opposite sign test (OST) [38] as the local search. A completely different approach, dealing with features in a diabetic clinical dataset, is proposed in [33]. Authors used genetic programming to generate new features from existing ones, without prior knowledge of the probability distribution. Sideris et al. proposed a novel, clustering-based (hierarchical clustering) feature extraction framework, using disease diagnostic information [34]. Their methodology produced clusters to be used as features for patient severity of condition and patient readmission risk prediction.

Finally, work on high-dimensional data was presented in [27]. Feature selection is a very challenging task, when performed in high-dimensional data such as genomic data. Cai et al. applied a feature selection method, called iterative sure independence screening (ISIS) [39] for gene profiles obtained from metagenome sequencing in Chinese/European cohorts, achieving 0.97/0.99 accuracy following selection of 48/24 meta-markers.

Prediction of DM

The second category deals with disease prediction and diagnosis [40–76]. Numerous algorithms and different approaches have been applied, such as traditional machine learning algorithms, ensemble learning approaches and association rule learning in order to achieve the best classification accuracy. Most noted among the aforementioned ones are the following:

Calisir and Dogantekin proposed LDA–MWSVM, a system for diabetes diagnosis [66]. The system performs feature extraction and reduction using the Linear Discriminant Analysis (LDA) method, followed by classification using the Morlet Wavelet Support Vector Machine (MWSVM) classifier. Gangji and Abadeh [65] proposed an Ant Colony-based classification system to extract a set of fuzzy rules, named FCS-ANTMINER, for diabetes diagnosis. In [68], authors dealt with glucose prediction as a multivariate regression problem utilizing Support Vector Regression (SVR). Agarwal et al. [48] utilized semi-automatically labeled training sets to create phenotype models via machine learning methods. In [76], authors proposed a fuzzy ontology-based Case-based reasoning (CBR) framework, mimicking expert thinking, further tested on diabetes diagnosis problems. In [58], authors performed an evaluation of Stream Mining Classifiers for Real-time Clinical Decision Support Systems.

With respect to high dimensional datasets, Razavian et al. [44] used a dataset containing 4.1 million individuals and 42K variables from administrative claims, pharmacy records, healthcare utilization, and laboratory results between 2005 and 2009, to build predictive models (based on logistic regression) for different onsets of T2D prediction.

A completely different study is presented in [40]. Authors built disease progression models, taking into account trajectories, i.e. the sequence of events leading to a state. When applied to Diabetes data, they identified a typical trajectory from hyperlipidemia (HLD) to hypertension (HTN), impaired fasting glucose (IFG), and T2D.

Ensemble approaches, which use multiple learning algorithms, have proven to be an effective way of improving classification accuracy. The specific approaches have also been used in DM prediction [50,52,53,69]. Anderson et al. used a Bayesian scoring algorithm to explore the model space [50]. In [52], authors proposed an ensemble framework with multi-layer classification, using enhanced bagging and optimized weighting, combining seven heterogeneous classifiers. In [53], authors used Rotation Forest (RF), a newly proposed ensemble algorithm, to combine 30 machine learning algorithms. Finally, Han et al. presented an ensemble learning approach, which turns the "black box" of SVM decisions into comprehensible and transparent rules [69].

Association Rules are mainly employed to identify associations between risk factors in an interpretable form [71–74]. In [71], authors applied association rules to detect combinations of variables or predictors frequently occurring together in diabetic patients. Simon et al. proposed Survival Association Rule (SAR) Mining [72], an extension to traditional Association Rule Mining, which can handle survival outcomes, make adjustment for confounders and incorporate dosage effects. In [73], authors reviewed four association rule set summarization techniques and proposed extensions, in order to deal with the large number of rules, mined from ARM applied to high dimensional EMR data. Finally, Batal et al. utilized temporal pattern mining for discovering predictive patterns in complex multivariate time series data, to improve performance of current classifiers [74].

Diabetic Complications

As mentioned above, the main pathophysiological feature in DM is hyperglycemia. In addition to normal glucose metabolism, prevention of complications due to elevated glucose levels is of great concern.

Generally, the harmful effects of hyperglycemia are divided into a) macrovascular complications, such as coronary artery disease, peripheral arterial disease, and stroke and, b) microvascular complications that include diabetic neuropathy, nephropathy, and retinopathy [77]. The direct and indirect effects of hyperglycemia are the main source of morbidity and mortality in both T1D and T2D. Large prospective clinical studies show a strong relationship between glycemia and diabetic microvascular complications in both T1D and T2D. Diabetic complications can also be classified according to their severity and time of onset. In these terms, acute diabetic complications include: diabetic ketoacidosis, hypoglycemia, diabetic coma, erectile dysfunction, respiratory infections and periodontal disease. Chronic diabetic complications include: heart failure, diabetic neuropathy, nephropathy, retinopathy, and diabetic foot. Moreover, both insulin resistance and hyperglycemia have been implicated in the pathogenesis of diabetic dyslipidemia. It is worth noting that DM complications are far less common and severe in people with well-controlled blood glucose levels. Many of those complications have been studied through machine learning and data mining applications [78–85,87–90,92,94–97].

In a more general aspect, Lagani et al. targeted several diabetic complications, such as cardiovascular diseases (CVD), hypoglycemia, ketoacidosis, microalbuminuria, proteinuria, neuropathy, and retinopathy [78,79]. In [78], in an effort to identify the smallest set of clinical parameters with the best predictive accuracy, involving the aforementioned diabetic complications, a set of predictive models was used that had been developed through data mining and machine learning approaches. In [80], authors used two distinct data sources (drug purchase and administrative information) to exploit temporal data mining techniques and improve risk stratification of diabetic complications.

In the case of nephropathy, Huang et al. employed a Decision Tree-based prediction tool that combines both genetic and clinical features in order to identify diabetic nephropathy in patients with T2D [81]. Leung et al. compared several machine learning methods that include partial least square regression, classification and regression tree, the C5.0 Decision Tree, Random Forest, naïve Bayes, neural networks and support vector machines [82]. The dataset used consists of both genetic (Single Nucleotide Polymorphisms — SNPs) and clinical data. Age,

age of diagnosis, systolic blood pressure and genetic polymorphisms of uteroglobin and lipid metabolism arose as the most efficient predictors.

Similarly, in the case of neuropathy, DuBrava et al. used Random Forest (RF) in order to select specific features targeting prediction of diabetic peripheral neuropathy (DPN) [83]. Based on relevance, the features chosen were Charlson Comorbidity Index score (100%), age (37.1%), number of pre-index procedures and services (29.7%), number of pre-index outpatient prescriptions (24.2%), number of pre-index outpatient visits (18.3%), number of pre-index laboratory visits (16.9%), number of pre-index outpatient office visits (12.1%), number of inpatient prescriptions (5.9%), and number of pain-related medication prescriptions (4.4%). The overall accuracy of the model reached 89%. The Database of the diabetes screening research initiative (DiScRi) [113] was used in [84,85] to predict Cardiovascular Autonomic Neuropathy (CAN). Staniery et al., used decision tree and optimal decision path finder (ODPF) to find the optimal sequences of Ewing tests to predict CAN, whereas Abawajy et al. used regression and meta-regression, in combination with the Ewing formula, to identify the classes in CAN, thus overcoming the problem of missing data.

Although Alzheimer›s disease is a chronic neurodegenerative disease, seemingly not related to DM, several studies support the fact DM and AD have a strong causal relationship [86]. Alzheimer›s disease is often referred to as “type 3” diabetes. In [87], authors delved into the relationship between DM and AD via semantic data mining. Following extensive analysis of several paper abstracts, they managed to identify genes related to both diseases. Efforts were also made to construct an interaction network in order to identify existing links (genes and molecules) in the network.

In [88], authors developed a predictive model exploiting data from two safety-net clinical trials that target comorbid depression, which could be considered as a diabetic complication, among patients with DM. In addition, in [89], authors tried to investigate the effectiveness of e-nose technology, using common classifiers, to predict single- and poly-microbial species targeted for diabetic foot infection. Rau et al. [90], also, developed a model to predict liver cancer within six years following T2D diagnosis. A dataset comprised of 2060 cases, was divided into two groups, encompassing patients a) diagnosed with liver cancer after diabetes, and b) with diabetes, but no liver cancer.

Heart-related abnormalities are considered as common diabetic complications [91]. It is worth noting that there›s a significant link between diabetes, heart disease, and stroke. In fact, two out of three people with diabetes die from heart disease or stroke, also called cardiovascular disease. In [92], researchers developed a hybrid approach, partially based on conditional random field classifier, to extract related information on heart disease risk factors from longitudinal unstructured EHRs.

Hypoglycemia, reflecting low blood sugar levels, arises mainly due to anti-diabetic treatment and has a great impact among DM patients [93]. Machine learning methods, such as Random Forest, support vector machines (SVM), k-nearest neighbor, and naïve Bayes, were used by Sudharsan B et al. [94] to predict hypoglycemia among patients with T2D, whereas support vector regression was used by Georga et al. [95] for the same reason. Moreover, a comparison of already published algorithms was reported by Jensen [96] in the same framework.

Intentional insulin treatment omission is an inappropriate compensatory behavior, occurring mainly in female patients with T1D, who omit or restrict their required insulin doses in order to lose weight. Although that does not occur as a diabetic complication but rather as a compensatory behavior, diagnosis of this underlying disorder is of great concern. In [97], authors used decision trees to analyze clinical and laboratory data for the prediction of intentional insulin omission for intentional weight loss.

Diabetic Retinopathy (DR) is an eye disease, occurring in people with either T1D or T2D. The longer a patient has diabetes the higher the risk of developing the specific pathophysiological condition. DR usually exhibits early warning signs and is characterized as a major diabetic complication [98]. DR can be divided into two main stages: a) NPDR (non-proliferative DR), and b) PDR (proliferative DR). Given the considerable impact of the current complication on patient lifestyle as well as society, numerous efforts have targeted accurate prediction of the disease onset in an effort to prevent progression.

Considering data mining and machine learning approaches, DR is the most studied field, mainly based on image processing techniques [100–112]. A comprehensive review on computational methods for diabetic retinopathy was published in 2013 [99]. Interestingly, in [100,101], data acquisition was also based on proteomic analyses. Specifically, Torok et al. developed a method, in which different types of data (results from tear fluid proteomics analysis and digital micro aneurysm detection on fundus images) were used as input in a Gradient Boosting Machine for DR screening, whereas Jin et al. performed comprehensive proteomics analysis to identify biomarkers for DR, concluding that a four protein biomarker panel (APO4, C7, CLU, and ITIH2) is capable of detecting early stages of the disease. Oh et al. [102] reported the first attempt in predicting DR using least absolute shrinkage and selection operator (LASSO) exploiting health record data. Moreover, Ibrahim et al. [103] used a data adaptive neuro fuzzy inference classifier to predict diabetes maculopathy. Roychowdhury et al. [104] targeted the degree of severity in DR, using a computer-aided screening system (DREAM) that analyzes fundus images with varying illumination and fields of view via machine learning approaches. A two phase method, Diabetic Fundus Image Recuperation (DFIR), was used in [105] for DR prediction. The first phase performs feature selection on digital retinal fundus images. The second phase employs a support vector machine for the prediction. A different aspect to the DR problem was investigated by Pires et al. [106]. In that case, a method for assessing the need for referral was developed, based on the identification of DR-related lesions in retinal images. Finally, Giancardo et al. proposed a methodology for Diabetic macular edema prediction, which is a common vision-threatening complication of DR [107]. Zhang et al. proposed a method for detecting DM and Non proliferative DR (early event) using tongue color, texture, and geometry features [110].

Drugs and Therapies

People with both types of diabetes need medication to help maintain normal blood sugar levels. The type of medication clearly depends on the type of diabetes. Insulin is the most common type of medication employed in T1D treatment and also used to treat T2D in some cases, depending on the severity of insulin depletion. At present, the majority of current therapies for T2D rely mainly on a number of approaches intending to reduce hyperglycemia. Such factors include sulfonylureas, metformins, PPAR-γ agonists (peroxisome proliferator-activated receptor-γ), α-glucosidase inhibitors, and others. Although diabetes constitutes a worldwide epidemic, with significant efforts targeting effective drug design and therapeutic protocols, most current therapies for this disease were developed in the absence of defined molecular targets or full delineation of the disease pathogenesis. Given the a) numerous side-effects of the present therapeutic protocols, and b) rapidly accruing knowledge on pathophysiological mechanisms, drug design and discovery stand as a great challenge in current research on diabetes. Intensive study of the mechanisms of action of older drugs has provided further validation of several recently identified drug targets. Further efforts in this direction are likely to be fruitful. In the era of post-genomic drug development, extracting and applying knowledge from biochemical, chemical, biological, and clinical data is one of the most interesting challenges facing the pharmaceutical industry. By the same token, data mining techniques can help a) recommend and improve effective medication, b) predict and suggest more personalized medications, c) design more effective blood glucose lowering factors, d) improve insulin planning and dosage, and e) implement drug administration in a more specific manner.

Sequential pattern mining techniques are used to mine patterns from data, where values are delivered in a sequence. Thus, such techniques are suitable in predicting the sequence of medications to be prescribed for a patient. Wright et al. used sequential pattern mining (CSPADE algorithm) to identify temporal relationships among medication prescriptions and finally predict the follow-up medication to be prescribed for a patient [114]. Also, Deja et al. used differential sequence patterns to imprint deviations observed in patient blood glucose levels and the amount of insulin dose to improve physician therapy recommendations [115].

In addition, to improve dosage planning, case-based reasoning was used to optimize the appropriate and effective dose of insulin in T1D [116]. By the same token, Karahoca and Tunga [117] used High Dimensional Model Representation (HDMR) to manage the drug dosage planning process in T2D. Moreover, taking into consideration patient behavior in relation to patient care, Namayanja and Janeja used clustering techniques to improve insulin treatment in T2D patients [118].

To search for novel anti-diabetic agents, the potency of inhibiting DPP4 was employed in [119], through decision tree classifier based on thirteen physicochemical properties, including molecular weight, total energy of a molecule, and topological polar surface area. A QSAR model was also used to assess flavonoid inhibitory effects on AR activity as a potent treatment for diabetes, using artificial neural networks [120]. In [121], a novel method was proposed, based on association rule mining, to discover relationships between statin (reductase inhibitors, medication for cardiovascular disease) use and diabetes. In addition, in [122], the study aimed at determining whether data mining methodologies could identify reproducible predictors of dapagliflozin-specific treatment response in the phase 3 clinical program dataset.

Liu et al. performed feature selection using wrapper and filter approaches on a 258 feature set in order to improve classification accuracy for medication recommendation in T2D using K-Nearest Neighbor [123].

Gastrointestinal surgery is considered as an alternative beneficial treatment for morbidly obese T2D patients. Authors in [124,125] targeted selection of markers for the prediction of successful T2D remission, following gastrointestinal surgery via artificial neural networks.

Postprandial hyperglycemia is considered as a global threat for both prediabetes and T2D. However, the current dietary methods for managing blood glucose levels exhibit limited efficacy. Zeevi et al. developed a machine learning algorithm that takes into account blood parameters, dietary habits, anthropometrics, physical activity, and gut microbiota to predict personalized postprandial glycemic response to real-life meals [126].

Genetic Background and Environment

Both type 1 and 2 diabetes as well as other rare forms of diabetes that are directly inherited, including MODY and diabetes due to mutations in mitochondrial DNA, are caused by a combination of genetic and environmental risk factors. Unlike some traits, diabetes does not seem to be inherited in a simple pattern. Undoubtedly, however, some people are born prone to developing diabetes more so than others. Several epidemiological patterns suggest that environmental factors contribute to the etiology of T1D. Interestingly, the recent elevated number of T1D incidents projects a changing global environment, which acts either as initiator and/or accelerator of beta cell autoimmunity rather than variation in the gene pool. Several genetic factors are involved in the development of the disease [127]. There is evidence that more than twenty regions of the genome are involved in the genetic susceptibility to T1D. The genes most strongly associated with T1D are located in the HLA region of chromosome 6 [128]. Similar to T1D, T2D has a strong genetic component. To date, more than 50 candidate genes for T2D have been investigated in various populations worldwide. Candidate genes are selected due to their interference with pancreatic beta cell function, insulin mode of action, glucose metabolism and/or other risk factors. It is a fact that advances in genotyping technology, over the past few years, have facilitated rapid progress in large-scale genetic studies. Identification of a large number of novel genetic variants increasing susceptibility diabetes and related traits opened up opportunities, not existing thus far, to associate this genetic information with clinical practice and possibly improve risk prediction. However, available data to date do not yet provide convincing evidence supporting use of genetic screening in the prediction of diabetes.

The human leukocyte antigen (HLA) system is a gene complex encoding the major histocompatibility complex (MHC) proteins in humans. HLA types are inherited and some of them are linked to autoimmune disorders and/or other diseases, including T1D diabetes. This fact has also been emphasized by recent genome-wide association studies. Zhao et al. attempted to reduce genetic association to practice through an HLA-based disease predictive model [129]. The authors managed to overcome the burden of low-predicting accuracy by using highly polymorphic genes as predictors. They proposed a methodology, which treats complex HLA genotypes as "objects", and built predictive models for T1D using eight HLA genes (HLA-DRB1, HLA-DRB3, HLA-DRB4, HLA-DRB5, HLA-DQA1, HLA-DQB1, HLA-DPA1, and HLA-DPB1). By the same token, authors in [130] analyzed 19,035 SNPs of 10,579 subjects and selected as few as three SNPs to predict HLA-DR/DQ types relevant to T1D.

Even pleiotropic genes have a strong impact on DM onset and progression. It›s a fact that pleiotropic genes cannot be easily associated with important diseases. To this end, Park et al. developed an association rule mining-based method to discover patterns of multiple phenotypic associations over 52 anthropometric and biochemical traits [131]. The discovered patterns were then used to identify genetic markers that can be associated with multivariate phenotypes.

It is worth mentioning that in [132], a meta-analysis study was conducted, where a collection of gene expression datasets of pancreatic beta-cells, conditioned in an environment resembling T1D induced apoptosis, such as exposure to pro-inflammatory cytokines, in order to identify relevant and differentially expressed genes. The specific genes were then characterized according to their function and prior literature-based information to build temporal regulatory networks. Moreover, biological experiments were carried out revealing that inhibition of two of the most relevant genes (RIPK2 and ELF3), previously unknown in T1D literature, have a certain impact on apoptosis.

On the other hand, Lee et al. used various classification algorithms, such as SVM and logistic regression, to predict T2D by employing 499 known SNPs from 87 T2D-related genes [133]. Finally, in [134], authors used support vector machines to predict tyrosine kinase ligand-receptor pairs from their amino acid sequences. More specifically, authors initially collected tyrosine kinase ligand-receptor pairs from the Database of Interacting Proteins (DIP) and UniProtKB, and after a filtering process, used them as a dataset for the assessment of predictive performance. Identification of the interacting partner of tyrosine kinase ligand-receptor, provides a deeper delineation of cellular-combined processes.

Health Care Management Systems

As mentioned above, the prevalence of diabetes for all age groups worldwide was estimated to be 2.8% in 2000 and 4.4% in 2030 [135]. The total number of people diagnosed with diabetes is projected to rise from 171 million in 2000 to 366 million by 2030. It is a grim fact that the majority of healthcare institutions in many countries spend billions of dollars on Diabetes health care. Given the impact of the disease, efforts are presently made to assess existing data in order to manage public health issues, such as hospitalization cost or medication.

In [136], authors presented a method to predict health care spending in ambulatory diabetes patients. In their method, authors used patterns extracted from health-related quality of life (HRQOL) inventories and electronic medical records and developed a hybrid approach based on Natural Language Processing and machine learning for the prediction models.

Nimmagadda et al. [137] developed a robust back-end application for web-based patient–doctor consultations and e-Health care, based on ontology-based multidimensional data warehousing and mining methodologies, while Renard et al. [138] developed DIABECOLUX, an algorithm for the prediction of treated T2D patients via health insurance claims, when no diagnosis code is available. Similarly, in [139], data from electronic health records and financial billing systems were used to produce integrated patient-based datasets. Mining of such data through probabilistic clustering methodologies allows assessment of the health and financial risk status, subsequently aiding in taking the appropriate proactive actions. Ultimately, Lee and Giaraud-Carrier [140] aimed at mining a huge collection of data, through association rules and clustering techniques, to support evidence-based medicine. Data were obtained from The National Health and Nutrition Examination Survey (NHANES), which is a program trying to assess health and nutritional status of adults and children in the United States.

DISCUSSION

In the present study, the recent literature was reviewed with respect to applications of machine learning and data mining methods in Diabetes research. The first sections describe briefly the two main research fields involved (machine learning, knowledge discovery in databases and Diabetes), pointing out the necessity of intelligent applications in improving the quality and effectiveness of decision making in DM.

Following creation of the assembled article collection (for methodology details vide supra), each article was categorized accordingly in one of the title groups (descending number of papers), thus covering to a great extent significant diabetes research fields, i.e. Biomarker Prediction and Diagnosis in DM, Diabetic Complications, Drugs and Therapies, Genetic Background and Environment, and Health Care Management. The current articles were published in several scientific journals that deal with distinct and wide fields of interest, including bioinformatics, biomedical engineering and diabetes. In Fig. 4, the scientific journals are presented in line with their appearance in the present collection, whereas Fig. 3 depicts the number of articles published per year.

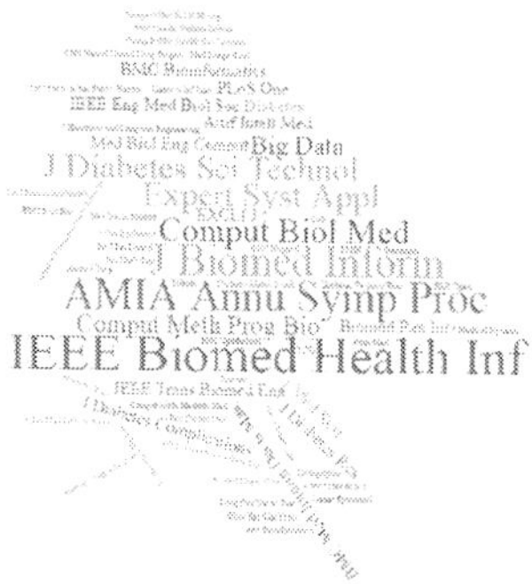

Figure 4: Distribution of articles in scientific journals.

The specific article categorization was carried out based on the content of the retrieved articles. The most popular category was the Biomarker Prediction and Diagnosis of DM, thematically revolving around efforts to discover and suggest novel biomarkers and finally predict key aspects of the disease, such as its onset. Since the undertaken research reflects a data-driven process, the arising gaps and limitations in machine learning research in DM are closely related to the availability of data. Clinical, diagnostic data and EHR are plentiful due to low cost of their retrieval, in contrast to other types of data, such as biological, which are more difficult and expensive to generate and therefore less available to the scientific community. That partially justifies the extensive research effort on specific topics, such as retinopathy. Moreover, there is complete lack of data concerning a) lifestyle and behavior, b) inheritance, and c) linkage with other pathophysiological conditions e.g. Alzheimer's disease.

Computational Insight Into Diabetes Research

When it comes to machine learning and data mining, significant conclusions are drawn through the present detailed account. It is worth mentioning that the vast majority of the reported articles enhanced classification accuracy, above 80%, in the prediction of DM. With regard to the prediction task itself, almost all of the common known classification algorithms have been employed. However, the most commonly used ones are SVM, ANN, and DT. It should be mentioned that SVM rises as the most successful algorithm in both biological and clinical datasets in DM. A great deal of articles (~ 85%) used the supervised learning approaches, i.e. in classification and regression tasks. In the remaining

15%, association rules were employed mainly to study associations between biomarkers. More specifically, concerning the part dealing with the evaluation task, in all reported research reports, the identified subsets of biomarkers (features) were evaluated through appropriate procedures, such as splitting the dataset into train and test set or via cross-validation. By analogy, the same approaches have been followed in DM prediction.

Worth emphasizing is the fact that in many studies, after the feature/biomarker selection, researchers have performed comparative analysis on different machine learning algorithms in order to assess their predictive performance and finally choose the most efficient one(s). To this end, this should be the baseline of practice in every study to be carried out, taking into account that several characteristics of the dataset, such as dimensionality, low number of instances compared to number of features or even the type of the dataset itself (genetic or clinical), can affect significantly the performance of the algorithm. Hence, an algorithm with the best performance in one dataset could easily have lower prediction accuracy compared to other algorithms in different datasets. Table 1 represents studies that compare more than five machine learning algorithms in various biological and clinical datasets. SVM exhibited the best performance with regard to classification accuracy or the Area Under the Curve (AUC). Moreover, many times in KDD, algorithms that produce interpretable results, are preferably used, although they are not necessarily state-of-the-art. The aforementioned fact explains, at least partially, the wide use of decision trees in the literature. The overall results project that a wide variety of algorithms and techniques are used in DM research. Obviously, different machine learning tasks are used in different scientific questions, such as prediction on DM or association among biomarkers. To this end, classification and regression techniques are used for prediction tasks, such as prediction of glucose levels and association rules in the case of dependencies between biomarkers. Interestingly, for each machine learning task, a variety of algorithms have been used in the literature. The reason behind that is likely the fact that the accuracy of an algorithm depends heavily on the type of data (dimensionality, origin and kind). Accordingly, a great effort in research relies on the preprocessing of data, such as feature selection and then various algorithms are applied to the processed data in order to identify the most successful one for the particular dataset.

Table 1: Comparison of different ML algorithms.

Publication	Type of DM	Type of data	No. of subjects	Compared algorithms	Validation method	Best accuracy
Cai et al. [27]	T2D	Gut microbiota	Dataset A: 344 Dataset B: 145	Logistic regression (LR), linear discriminant analysis (LDA), naïve Bayes (NB) and support vector machine (SVM)	10-fold cross-validation	SVM on several different experiments
Malik et al. [46]	Both types (hyperglycemia)	Electrochemical measurements of saliva	175	Logistic regression (LR), support vector machine (SVM) and artificial neural network (ANN)	3-fold cross-validation	SVM ACC = 84.09
Farran et al. [60]	T2D	Demographic, anthropometric, vital signs, diagnostic and clinical laboratory measurements	10,632	Logistic regression (LR), k-nearest neighbors (k-NN), multifactor dimensionality reduction (MDR) support vector machines (SVM)	5-fold cross-validation	SVM ACC = 81.3

Mani et al. [61]	T2D	Demographic, clinical lab values	2280 distributed in three datasets	Gaussian Naïve Bayes (NB), Logistic Regression (LR), K-nearest neighbor (k-NN, CART, Random Forests (RF), Support Vector Machine (SVM)	5-fold cross-validation	RF AUC = 0.803/ 0.807/0.877
Tapak et al. [141]	Nonspecific	Demographic, anthropometric, diagnostic and clinical laboratory measurements	6500	Artificial neural networks (ANN), support vector machines (SVM), fuzzy c-mean, Random Forests (RF)	10-fold cross-validation	SVM ACC = 0.986 AUC = 0.979

Furthermore, it is imperative for machine learning studies that a dataset be sufficiently large for the algorithm to be trained appropriately. Although biomedical sciences have entered the era of big data for several reasons, such as low cost of next generation sequencing or unified EHRs, datasets with great variability in size are very common in DM research. In that respect, what should be stressed out is the danger of a) producing low quality results, and b) concomitantly having the entire KDD process finally extract low quality of knowledge, when a small amount of data is employed.

Computational Interfacing With Diabetes Mellitus

Potential gains of early detection of a disease, in this case DM, in addition to the assessment of possible risk factors, include a) significant prolongation and quality of life, pertaining to the reduction of severity and frequency of a disease state and/or prevention and delay of its complications, and b) reduction of health care cost, as a consequence of reduced care linked to hospitalization of patients. In this context, data mining and machine learning arise as a key process providing insight into possible relationships among molecules and conditions such as gene–gene, protein–protein, drug–drug, drug–disease or gene–disease, etc.

From the perspective of DM, although there are several types of diabetes, the overall results suggest that the articles reviewed refer to T1D and T2D, with T2D representing the majority of the articles. A few articles refer to prediabetes and only one pertains to the metabolic syndrome, which is a term for metabolism-related pathophysiology. The types of data used in each case of the present collection were either clinical, genetic, electrochemical, chemical or medical. Only a few articles used clinical data in combination with genetic data. In addition, it is worth mentioning that the vast majority of the articles reviewed handled only clinical datasets. When it comes to prediction, the main biomarkers used involve anthropometric parameters, demographic characteristics, known risk factors, medical and drug history data, laboratory measurements, and epidemiological data. The most common biomarker seems to be blood glucose levels (HbA1c), as expected, since its detection is the basic step toward diagnosis and classification of a candidate diabetic patient.

With regard to DM treatment, the articles associated with drugs and therapy cover several fields of interest that include a) medication prescriptions, b) dosage planning with emphasis on insulin administration, c) potential side-effects of medications non-related to the disease (e.g. statins), and d) prediction of personalized glycemic response following anti-diabetic medication. Only Shoombautong et al., in [119], deal with the discovery of novel anti-diabetic agents. Therefore, to our knowledge there is much work ahead to be done on drug and therapeutic protocol design as far as evaluation and data mining on already known blood glucose lowering factors, such as metformin.

Concerning the genetic background in DM and environmental factors affecting the onset and progression of the disease, it is worth noting that the present account presents an evident gap in research on diabetes with respect to data mining and machine learning. The articles reviewed employ the HLA gene complex, in relation to T1D, whereas the rest of them attempt to predict associations of pleiotropic genes with DM. Interestingly, Lopes et al. tried to associate two known genes with DM, following wet lab validation of the extracted information [132]. Finally, although SNPs are one of the most common genetic markers in various research fields, in the present study only two articles utilized SNPs to predict DM. As more genes involved in the pathogenesis of diabetes are gradually identified, it will become easier to gain deeper understanding of the mechanisms responsible for the disease development and progression. That can lead to new insights into the genetic epidemiology of diabetes and nature of gene–gene and gene–environment interactions.

Finally, Diabetic complications covered in the present study include nephropathy, Alzheimer's disease, diabetic foot, liver cancer, hypoglycemic events, heart disease, depression, and retinopathy. The majority of the articles deal with retinopathy. One plausible explanation, apart from the impact of the disease, could be the availability of data resources from routine clinical practice that allow information extraction.

CONCLUSIONS

In this study, a systematic effort was made to identify and review machine learning and data mining approaches applied on DM research. DM is rapidly emerging as one of the greatest global health challenges of the 21st century. To date, there is a significant work carried out in almost all aspects of DM research and especially biomarker identification and prediction-diagnosis. The advent of biotechnology, with the vast amount of data produced, along with the increasing amount of EHRs is expected to give rise to further in-depth exploration toward diagnosis, etiopathophysiology and treatment of DM through employment of machine learning and data mining techniques in enriched datasets that include clinical and biological information.

ACKNOWLEDGEMENTS

This work has been partially supported by Horizon 2020 Framework Programme of the European Union under grant agreement 644906, the AEGLE project.

REFERENCES

1. Marx V. Biology: the big challenges of big data. Nature Jun 13 2013;498(7453): 255–60. http://dx.doi.org/10.1038/498255a.
2. Mattmann CA. Computing: a vision for data science. Nature Jan 24 2013; 493(7433):473–5. http://dx.doi.org/10.1038/493473a.
3. Wilson RA, Keil FC. The MIT encyclopaedia of the cognitive sciences. MIT Press; 1999.
4. Mitchell T. Machine learning. McGraw Hill0-07-042807-7; 1997 2.
5. Fayyad U, Piatetsky-Shapiro G, Smyth P. From data mining to knowledge discovery in databases. AI Mag 1996;17:37–54.
6. Russell, Stuart; Norvig, Peter (2003) [1995]. Artificial Intelligence: A Modern Approach (2nd Ed.). Prentice Hall. ISBN 978-0137903955.
7. Agrawal R, Imielinski T, Swami A. Mining association rules between sets of items in large databases. Proceedings of the ACM SIGMOD conference on management of data; 1993. p. 207–16.
8. Agrawal R, Srikant R. Fast algorithms for mining association rules in large databases. Proceedings of the 20th International Conference on Very Large Databases; 1994. p. 478–99.
9. Kavakiotis I, Tzanis G, Vlahavas I. Mining frequent patterns and association rules from biological data. In: Elloumi M, Zomaya AY, editors. Biological knowledge discovery handbook: preprocessing, mining and postprocessing of biological data. Wiley Book series

on bioinformatics: computational techniques and engineeringNew Jersey, USA: Wiley-Blackwell, John Wiley & Sons Ltd.; 2014 [Publish.].

10. Han J, Kamber M, Pei J. Data mining: concepts and techniques. The Morgan Kaufmann series in data management systems; 2011.
11. Alpaydin E. Introduction to machine learning. Cambridge Massachusetts London England: The MIT Press; 2004.
12. Guyon I, Elisseeff A. An introduction to variable and feature selection. J Mach Learn Res 2003;3:1157–82.
13. Witten IH, Frank E, Hall MA. Data mining: practical machine learning tools and techniques. 3rd ed. Burlington, MA: Morgan Kaufmann; 2011.
14. American Diabetes Association. Diagnosis and classification of diabetes mellitus. Diabetes Care 2009;32(Suppl. 1):S62–7.
15. Cox EM, Elelman D. Test for screening and diagnosis of type 2 diabetes. Clin Diabetes 2009;4(27):132–8.
16. Krentz AJ, Bailey CJ. Oral antidiabetic agents: current role in type 2 diabetes mellitus. Drugs 2005;65(3):385–411.
17. Tsave O, Halevas E, Yavropoulou MP, Kosmidis Papadimitriou A, Yovos JG, Hatzidimitriou A, et al. Structure-specific adipogenic capacity of novel, welldefined ternary Zn(II)-Schiff base materials. Biomolecular correlations in zincinduced differentiation of 3T3-L1 pre-adipocytes to adipocytes. J Inorg Biochem Nov 2015;152:123–37. http://dx.doi.org/10.1016/j.jinorgbio.2015.08.014 [Epub 2015 Aug 11].
18. Halevas E, Tsave O, Yavropoulou MP, Hatzidimitriou A, Yovos JG, Psycharis V, et al. Design, synthesis and characterization of novel binary V(V)-Schiff base materials linked with insulin-mimetic vanadium-induced differentiation of 3T3-L1 fibroblasts to adipocytes. Structure–function correlations at the molecular level. J Inorg Biochem Jun 2015;147:99–115. http://dx.doi.org/10.1016/j.jinorgbio.2015. 03.009 [Epub 2015 Mar 26].
19. Tsave O, Yavropoulou MP, Kafantari M, Gabriel C, Yovos JG, Salifoglou A. The adipogenic potential of Cr(III). A molecular approach exemplifying metalinduced enhancement of insulin

mimesis in diabetes mellitus II. J Inorg Biochem Oct 2016;163:323–31.

20. Sakurai H, Kojima Y, Yoshikawa Y, Kawabe K, Yasui H. Antidiabetic vanadium(IV) and zinc(II) complexes review article coordination. Chem Rev March 2002; 226(1–2):187–98.
21. "Records in DBLP". Statistics. DBLP. Retrieved 2016–07-16; 2016
22. Després J-P, Lemieux I. Abdominal obesity and metabolic syndrome. Nature December 14 2006;444:881–7 [Published online 13 December 2006].
23. Caveney EJ, Cohen OJ. Diabetes and biomarkers. J Diabetes Sci Technol Jan 2011; 5(1):192–7.
24. Jelinek HF, Stranieri A, Yatsko A, Venkatraman S. Data analytics identify glycated haemoglobin co-markers for type 2 diabetes mellitus diagnosis. Comput Biol Med Aug 1 2016;75:90–7. http://dx.doi.org/10.1016/j.compbiomed.2016.05.005 [Epub 2016 May 13].
25. Bagherzadeh-Khiabani F, Ramezankhani A, Azizi F, Hadaegh F, Steyerberg EW, Khalili D. A tutorial on variable selection for clinical prediction models: feature selection methods in data mining could improve the results. J Clin Epidemiol 2016 Mar;71:76–85. http://dx.doi.org/10.1016/j.jclinepi.2015.10.002 [Epub 2015 Oct 22].
26. Wang KJ, Adrian AM, Chen KH, Wang KM. An improved electromagnetism-like mechanism algorithm and its application to the prediction of diabetes mellitus. J Biomed Inform Apr 2015;54:220–9. http://dx.doi.org/10.1016/j.jbi.2015.02.001 [Epub 2015 Feb 10].
27. Cai L, Wu H, Li D, Zhou K, Zou F. Type 2 diabetes biomarkers of human gut microbiota selected via iterative sure independent screening method. PLoS One Oct 19 2015;10(10):e0140827. http://dx.doi.org/10.1371/journal.pone.0140827 [eCollection 2015].
28. Georga EI, Protopappas VC, Polyzos D, Fotiadis DI. Evaluation of short-term predictors of glucose concentration in type 1 diabetes combining feature ranking with regression models. Med Biol Eng Comput Dec 2015;53(12):1305–18. http://dx.doi. org/10.1007/s11517–015–1263-1 [Epub 2015 Mar 15].

29. Lee BJ, Kim JY, Lee BJ, Kim JY. Identification of type 2 diabetes risk factors using phenotypes consisting of anthropometry and triglycerides based on machine learning. IEEE J Biomed Health Inform Jan 2016;20(1):39–46. http://dx.doi.org/10.1109/JBHI.2015.2396520 [Epub 2015 Feb 6]. Marling CR, Struble NW, Bunescu RC, Shubrook JH, Schwartz FL. A consensus perceived glycemic variability metric. J Diabetes Sci Technol Jul 1 2013;7(4):871–9.
30. Huang JH, He RH, Yi LZ, Xie HL, Cao DS, Liang YZ. Exploring the relationship between 5′AMP-activated protein kinase and markers related to type 2 diabetes mellitus. Talanta Jun 15 2013;110:1–7. http://dx.doi.org/10.1016/j.talanta.2013.03. 039 [Epub 2013 Mar 22].
31. Worachartcheewan A, Nantasenamat C, Isarankura-Na-Ayudhya C, Prachayasittikul V. Quantitative population–health relationship (QPHR) for assessing metabolic syndrome. EXCLI J Jun 26 2013;12:569–83 [eCollection 2013].
32. Aslam MW, Zhu Z, Nandi AK. Feature generation using genetic programming with comparative partner selection for diabetes classification. Expert Syst Appl 2013; 40(13):5402–12.
33. Sideris C, Pourhomayoun M, Kalantarian H, Sarrafzadeh M. A flexible data-driven comorbidity feature extraction framework. Comput Biol Med Jun 1 2016;73: 165–72. http://dx.doi.org/10.1016/j.compbiomed.2016.04.014 [Epub 2016 Apr 20].
34. Breiman L. Random forests. Mach Learn 2001;45(1):5–32. http://dx.doi.org/10. 1023/A:1010933404324.
35. Robnik-Sikonja M, Kononenko I. Theoretical and empirical analysis of ReliefF and RReliefF. Mach Learn 2003;53(1–2):23–69. http://dx.doi.org/10.1023/ A: 1025667309714.
36. Cover TM, Hart PE. Nearest neighbor pattern classification. IEEE Trans Inf Theory 1967;IT-13(1):21–7.
37. Chen LF, Su CT, Chen KH. An improved particle swarm optimization for feature selection. Intell Data Anal 2012;16(2):167–82.
38. Fan J, Lv J. Sure independence screening for ultrahigh dimensional feature space. J R Stat Soc Series B Stat Methodology 2008;70(5):849–911. http://dx.doi.org/10. 1111/j.1467-9868. 2008.00674.x.

39. Oh W, Kim E, Castro MR, Caraballo PJ, Kumar V, Steinbach MS, et al. Type 2 diabetes mellitus trajectories and associated risks. Big Data Mar 1 2016;4(1):25–30.

40. Worachartcheewan A, Nantasenamat C, Prasertsrithong P, Amranan J, Monnor T, Chaisatit T, et al. Machine learning approaches for discerning intercorrelation of hematological parameters and glucose level for identification of diabetes mellitus. EXCLI J Oct 21 2013;12:885–93 [eCollection 2013].

41. Worachartcheewan A, Shoombuatong W, Pidetcha P, Nopnithipat W, Prachayasittikul V, Nantasenamat C. Predicting metabolic syndrome using the random forest method. ScientificWorldJournal 2015;2015:581501. http://dx.doi.org/ 10.1155/2015/581501 [Epub 2015 Jul 28].

42. Habibi S, Ahmadi M, Alizadeh S. Type 2 diabetes mellitus screening and risk factors using decision tree: results of data mining. Glob J Health Sci Mar 18 2015;7(5): 304–10. http://dx.doi.org/10.5539/gjhs.v7n5p304.

43. Razavian N, Blecker S, Schmidt AM, Smith-McLallen A, Nigam S, Sontag D. Population-level prediction of type 2 diabetes from claims data and analysis of risk factors. Big Data Dec 2015;3(4):277–87. http://dx.doi.org/10.1089/big.2015. 0020.

44. Meng XH, Huang YX, Rao DP, Zhang Q, Liu Q. Comparison of three data mining models for predicting diabetes or prediabetes by risk factors. Kaohsiung J Med Sci Feb 2013;29(2):93–9. http://dx.doi.org/10.1016/j.kjms.2012.08.016 [Epub 2012 Oct 16].

45. Malik S, Khadgawat R, Anand S, Gupta S. Non-invasive detection of fasting blood glucose level via electrochemical measurement of saliva. Springerplus May 23 2016;5(1):701. http://dx.doi.org/10.1186/s40064-016-2339-6 [eCollection 2016].

46. Allalou A, Nalla A, Prentice KJ, Liu Y, Zhang M, Dai FF, et al. A predictive metabolic signature for the transition from gestational diabetes to type 2 diabetes. Diabetes Jun 23 2016 [pii: db151720. [Epub ahead of print]].

47. Agarwal V, Podchiyska T, Banda JM, Goel V, Leung TI, Minty EP, et al. Learning statistical models of phenotypes using noisy labeled

training data. J Am Med Inform Assoc May 12 2016. http://dx.doi.org/10.1093/jamia/ocw028 [pii: ocw028 [Epub ahead of print]].

48. Hoyt R, Linnville S, Thaler S, Moore J. Digital family history data mining with neural networks: a pilot study. Perspect Health Inf Manag Jan 1 2016;13:1c [eCollection 2016].

49. Anderson JP, Parikh JR, Shenfeld DK, Ivanov V, Marks C, Church BW, et al. Reverse engineering and evaluation of prediction models for progression to type 2 diabetes: an application of machine learning using electronic health records. J Diabetes Sci Technol Dec 20 2015;10(1):6–18. http://dx.doi.org/10.1177/1932296815620200.

50. Anderson AE, Kerr WT, Thames A, Li T, Xiao J, Cohen MS. Electronic health record phenotyping improves detection and screening of type 2 diabetes in the general United States population: a cross-sectional, unselected, retrospective study. J Biomed Inform Apr 2016;60:162–8. http://dx.doi.org/10.1016/j.jbi.2015.12.006 [Epub 2015 Dec 17].

51. Bashir S, Qamar U, Khan FH. IntelliHealth: a medical decision support application using a novel weighted multi-layer classifier ensemble framework. J Biomed Inform Feb 2016;59:185–200. http://dx.doi.org/10.1016/j.jbi.2015.12.001 [Epub 2015 Dec 15].

52. Ozcift A, Gulten A. Classifier ensemble construction with rotation forest to improve medical diagnosis performance of machine learning algorithms. Comput Methods Programs Biomed 2011 Dec;104(3):443–51. http://dx.doi.org/10.1016/j.cmpb. 2011.03.018 [Epub 2011 Apr 30].

53. Ramezankhani A, Pournik O, Shahrabi J, Azizi F, Hadaegh F, Khalili D. The impact of oversampling with SMOTE on the performance of 3 classifiers in prediction of type 2 diabetes. Med Decis Making 2016 Jan;36(1):137–44. http://dx.doi.org/10.1177/ 0272989X14560647 [Epub 2014 Dec 1].

54. Choi SB, Kim WJ, Yoo TK, Park JS, Chung JW, Lee YH, et al. Screening for prediabetes using machine learning models. Comput Math Methods Med 2014;2014:618976. http://dx.doi.org/10.1155/2014/618976 [Epub 2014 Jul 16].

55. Belciug S, Gorunescu F. Error-correction learning for artificial neural networks using the Bayesian paradigm. Application to automated

medical diagnosis. J Biomed Inform 2014 Dec;52:329–37. http://dx.doi.org/10.1016/j.jbi.2014.07.013 [Epub 2014 Jul 21].

56. Lee BJ, Ku B, Nam J, Pham DD, Kim JY. Prediction of fasting plasma glucose status using anthropometric measures for diagnosing type 2 diabetes. IEEE J Biomed Health Inform Mar 2014;18(2):555–61. http://dx.doi.org/10.1109/JBHI.2013. 2264509.

57. Fong S, Zhang Y, Fiaidhi J, Mohammed O, Mohammed S. Evaluation of stream mining classifiers for real-time clinical decision support system: a case study of blood glucose prediction in diabetes therapy. Biomed Res Int 2013;2013:274193. http:// dx.doi.org/10.1155/2013/274193 [Epub 2013 Sep 19].

58. Ozery-Flato M, Parush N, El-Hay T, Visockienė Z, Ryliškytė L, Badarienė J, et al. Predictive models for type 2 diabetes onset in middle-aged subjects with the metabolic syndrome. Diabetol Metab Syndr Jul 15 2013;5(1):36. http://dx.doi.org/10.1186/ 1758–5996–5-36.

59. Farran B, Channanath AM, Behbehani K, Thanaraj TA. Predictive models to assess risk of type 2 diabetes, hypertension and comorbidity: machine-learning algorithms and validation using national health data from Kuwait—a cohort study. BMJ Open May 14 2013;3. http://dx.doi.org/10.1136/bmjopen-2012-002457 [pii: e002457].

60. Mani S, Chen Y, Elasy T, Clayton W, Denny J. Type 2 diabetes risk forecasting from EMR data using machine learning. AMIA Annu Symp Proc 2012;2012:606–15 [Epub 2012 Nov 3].

61. Shankaracharya, Odedra D, Samanta S, Vidyarthi AS. Computational intelligencebased diagnosis tool for the detection of prediabetes and type 2 diabetes in India. Rev Diabet Stud Spring 2012;9(1):55–62. http://dx.doi.org/10.1900/RDS.2012.9. 55 [Epub 2012 May 10].

62. Chikh MA, Saidi M, Settouti N. Diagnosis of diabetes diseases using an Artificial Immune Recognition System2 (AIRS2) with fuzzy K-nearest neighbor. J Med Syst Oct 2012;36(5):2721–9. http://dx.doi.org/10.1007/s10916-011-9748-4 [Epub 2011 Jun 22].

63. Malley JD, Kruppa J, Dasgupta A, Malley KG, Ziegler A. Probability machines: consistent probability estimation using nonparametric

learning machines. Methods Inf Med 2012;51(1):74–81. http:// dx.doi.org/10.3414/ME00–01-0052 [Epub 2011 Sep 14].

64. Ganji MF, Abadeh MS. A fuzzy classification system based on ant colony optimization for diabetes disease diagnosis. Expert Syst Appl 2011;38(12):14650–9.
65. Çalisir D, Dogantekin E. An automatic diabetes diagnosis system based on LDAWavelet Support Vector Machine Classifier. Expert Syst Appl 2011;38(7):8311–5.
66. Robertson G, Lehmann ED, Sandham WA, Hamilton DJ. Blood glucose prediction using artificial neural networks trained with the AIDA diabetes simulator: a proof-of-concept pilot study. J Electr Comput Eng 2011;2011: 681786:1–681786:11.
67. Georga EI, Protopappas VC, Ardigò D, Marina M, Zavaroni I, Polyzos D, et al. Multivariate prediction of subcutaneous glucose concentration in type 1 diabetes patients based on support vector regression. IEEE J Biomed Health Inform 2013; 17(1):71–81.
68. Han L, Luo S, Yu J, Pan L, Chen S. Rule extraction from support vector machines using ensemble learning approach: an application for diagnosis of diabetes. IEEE J Biomed Health Inform 2015;19(2):728–34.
69. Gregori D, Petrinco M, Bo S, Rosato R, Pagano E, Berchialla P, et al. Using data mining techniques in monitoring diabetes care. The simpler the better? J Med Syst 2011; 35(2):277–81.
70. Ramezankhani A, Pournik O, Shahrabi J, Azizi F, Hadaegh F. An application of association rule mining to extract risk pattern for type 2 diabetes using Tehran lipid and glucose study database. Int J Endocrinol Metab Apr 30 2015;13(2):e25389. http:// dx.doi. org/10.5812/ijem.25389 [eCollection 2015].
71. Simon GJ, Schrom J, Castro MR, Li PW, Caraballo PJ. Survival association rule mining towards type 2 diabetes risk assessment. AMIA Annu Symp Proc Nov 16 2013; 2013:1293–302 [eCollection 2013].
72. Simon GJ, Caraballo PJ, Therneau TM, Cha SS, Castro MR, Li PW. Extending association rule summarization techniques to assess risk of diabetes mellitus. IEEE Trans Knowl Data Eng 2015;27(1):130–41.

73. Batal I, Fradkin D, Harrison J, Moerchen F, Hauskrecht M. Mining recent temporal patterns for event detection in multivariate time series data. KDD. 2012; 2012. p. 280–8.

74. Beloufa F, Chikh MA. Design of fuzzy classifier for diabetes disease using modified artificial bee colony algorithm. Comput Methods Programs Biomed Oct 2013; 112(1):92–103. http://dx.doi.org/10.1016/j.cmpb.2013.07.009 [Epub 2013 Aug 7].

75. El-Sappagh S, Elmogy M, Riad AM. A fuzzy-ontology-oriented case-based reasoning framework for semantic diabetes diagnosis. Artif Intell Med Nov 2015;65(3): 179–208. http://dx.doi.org/10.1016/j.artmed.2015.08.003 [Epub 2015 Aug 14].

76. Cade WT. Diabetes-related microvascular and macrovascular diseases in the physical therapy setting. Phys Ther Nov 2008;88(11):1322–35.

77. Lagani V, Chiarugi F, Thomson S, Fursse J, Lakasing E, Jones RW, et al. Development and validation of risk assessment models for diabetes-related complications based on the DCCT/EDIC data. J Diabetes Complications May-Jun 2015;29(4):479–87. http://dx.doi.org/10.1016/j.jdiacomp.2015.03.001 [Epub 2015 Mar 6].

78. Lagani V, Chiarugi F, Manousos D, Verma V, Fursse J, Marias K, et al. Realization of a service for the long-term risk assessment of diabetes-related complications. J Diabetes Complications Jul 2015;29(5):691–8. http://dx.doi.org/10.1016/j.jdiacomp.2015.03.011 [Epub 2015 Mar 25].

79. Sacchi L, Dagliati A, Segagni D, Leporati P, Chiovato L, Bellazzi R. Improving riskstratification of diabetes complications using temporal data mining. Conf Proc IEEE Eng Med Biol Soc Aug 2015;2015:2131–4. http://dx.doi.org/10.1109/EMBC.2015.7318810.

80. Huang G-M, Huang K-Y, Lee T-Y, Weng J. An interpretable rule-based diagnostic classification of diabetic nephropathy among type 2 diabetes patients. BMC Bioinforma 2015;16(S-1):S5.

81. Leung RK, Wang Y, Ma RC, Luk AO, Lam V, Ng M, et al. Using a multi-staged strategy based on machine learning and mathematical modeling to predict genotype–phenotype risk patterns in diabetic kidney disease: a prospective case–control cohort analysis. BMC

Nephrol Jul 23 2013;14:162. http://dx.doi.org/10.1186/1471-2369-14-162.

82. DuBrava S, Mardekian J, Sadosky A, Bienen EJ, Parsons B, Hopps M, et al. Using random forest models to identify correlates of a diabetic peripheral neuropathy diagnosis from electronic health record data. Pain Med May 31 2016 [pii: pnw096. [Epub ahead of print]].

83. Stranieri A, Abawajy J, Kelarev A, Huda S, Chowdhury M, Jelinek HF. An approach for Ewing test selection to support the clinical assessment of cardiac autonomic neuropathy. Artif Intell Med Jul 2013;58(3):185–93. http://dx.doi.org/10.1016/j.artmed.2013.04.007 [Epub 2013 Jun 13].

84. Abawajy J, Kelarev A, Chowdhury M, Stranieri A, Jelinek HF. Predicting cardiac autonomic neuropathy category for diabetic data with missing values. Comput Biol Med Oct 2013;43(10):1328–33. http://dx.doi.org/10.1016/j.compbiomed.2013.07.002 [Epub 2013 Jul 12]. de la Monte SM, Wands JR. Alzheimer›s disease is type 3 diabetes—evidence reviewed. J Diabetes Sci Technol Nov 2008;2(6):1101–13.

85. Narasimhan K, Govindasamy M, Gauthaman K, Kamal MA, Abuzenadeh AM, AlQahtani M, et al. Diabetes of the brain: computational approaches and interventional strategies. CNS Neurol Disord Drug Targets Apr 2014;13(3):408–17.

86. Jin H, Wu S, Di Capua P. Development of a clinical forecasting model to predict comorbid depression among diabetes patients and an application in depression screening policy making. Prev Chronic Dis Sep 3 2015;12:E142. http://dx.doi.org/ 10.5888/pcd12.150047.

87. Yusuf N, Zakaria A, Omar MI, Shakaff AY, Masnan MJ, Kamarudin LM, et al. In-vitro diagnosis of single and poly microbial species targeted for diabetic foot infection using e-nose technology. BMC Bioinforma May 14 2015;16:158. http://dx.doi.org/ 10.1186/s12859-015-0601-5.

88. Rau HH, Hsu CY, Lin YA, Atique S, Fuad A, Wei LM, et al. Development of a webbased liver cancer prediction model for type II diabetes patients by using an artifi- cial neural network. Comput

Methods Programs Biomed Mar 2016;125:58–65. http://dx.doi.org/10.1016/j.cmpb.2015.11.009 [Epub 2015 Nov 27].

89. Patterson CC. Mortality from heart disease in a cohort of 23,000 patients with insulin-treated diabetes. Diabetologia 2003;46:760–5.

90. Jonnagaddala J, Liaw ST, Ray P, Kumar M, Dai HJ, Hsu CY. Identification and progression of heart disease risk factors in diabetic patients from longitudinal electronic health records. Biomed Res Int 2015;2015:636371. http://dx.doi.org/10.1155/ 2015/636371 [Epub 2015 Aug 25].

91. Cryer PE, Davis SN, Shamoon H. Hypoglycemia in diabetes. Diabetes Care 2003 [Am Diabetes Association].

92. Sudharsan B, Peeples M, Shomali M. Hypoglycemia prediction using machine learning models for patients with type 2 diabetes. J Diabetes Sci Technol Jan 2015;9(1):86–90. http://dx.doi.org/10.1177/1932296814554260 [Epub 2014 Oct 14].

93. Georga EI, Protopappas VC, Ardigò D, Polyzos D, Fotiadis DI. A glucose model based on support vector regression for the prediction of hypoglycemic events under freeliving conditions. Diabetes Technol Ther Aug 2013;15(8):634–43. http://dx.doi. org/10.1089/dia.2012.0285 [Epub 2013 Jul 13].

94. Jensen MH, Mahmoudi Z, Christensen TF, Tarnow L, Seto E, Johansen MD, et al. Evaluation of an algorithm for retrospective hypoglycemia detection using professional continuous glucose monitoring data. J Diabetes Sci Technol Jan 1 2014;8(1):117–22 [Epub ahead of print].

95. Pinhas-Hamiel O, Hamiel U, Greenfield Y, Boyko V, Graph-Barel C, Rachmiel M, et al. Detecting intentional insulin omission for weight loss in girls with type 1 diabetes mellitus. Int J Eat Disord Dec 2013;46(8):819–25. http://dx.doi.org/10.1002/ eat.22138 [Epub 2013 May 15].

96. Tapp RJ, Shaw JE, Harper CA, et al. The prevalence of and factors associated with diabetic retinopathy in the Australian population. Diabetes Care June 2003;26(6): 1731–7.

97. Li B, Li HK. Automated analysis of diabetic retinopathy images: principles, recent developments, and emerging trends. Curr Diab

Rep Aug 2013;13(4):453–9. http://dx.doi.org/10.1007/s11892–013–0393-9.

98. Torok Z, Peto T, Csosz E, Tukacs E, Molnar AM, Berta A, et al. Combined methods for diabetic retinopathy screening, using retina photographs and tear fluid proteomics biomarkers. J Diabetes Res 2015; 2015:623619. http://dx.doi.org/10.1155/2015/ 623619 [Epub 2015 Jun 29].

99. Jin J, Min H, Kim SJ, Oh S, Kim K, Yu HG, et al. Development of diagnostic biomarkers for detecting diabetic retinopathy at early stages using quantitative Proteomics.J. Diabetes Res 2016; 2016:6571976. http://dx.doi.org/10.1155/2016/ 6571976 [Epub 2015 Nov 9].

100. Oh E, Yoo TK, Park EC. Diabetic retinopathy risk prediction for fundus examination using sparse learning: a cross-sectional study. BMC Med Inform Decis Mak Sep 13 2013;13:106. http://dx.doi.org/10.1186/1472–6947–13-106.

101. Ibrahim S, Chowriappa P, Dua S, Acharya UR, Noronha K, Bhandary S, et al. Classi- fication of diabetes maculopathy images using data-adaptive neuro-fuzzy inference classifier. Med Biol Eng Comput Dec 2015;53(12):1345–60. http://dx.doi. org/10.1007/s11517–015–1329-0 [Epub 2015 Jun 25].

102. Roychowdhury S, Koozekanani DD, Parhi KK. DREAM: diabetic retinopathy analysis using machine learning. IEEE J Biomed Health Inform Sep 2014;18(5):1717–28. http://dx.doi.org/10.1109/ JBHI.2013.2294635.

103. Krishnamoorthy S, Alli P. A novel image recuperation approach for diagnosing and ranking retinopathy disease level using diabetic fundus image. PLoS One May 14 2015;10(5):e0125542. http:// dx.doi.org/10.1371/journal.pone.0125542 [eCollection 2015].

104. Pires R, Jelinek HF, Wainer J, Goldenstein S, Valle E, Rocha A. Assessing the need for referral in automatic diabetic retinopathy detection. IEEE Trans Biomed Eng Dec 2013;60(12):3391–8. http:// dx.doi.org/10.1109/TBME.2013.2278845 [Epub 2013 Aug 16].

105. Giancardo L, Meriaudeau F, Karnowski TP, Li Y, Garg S, Tobin Jr KW, et al. Exudatebased diabetic macular edema detection in fundus images using publicly available datasets. Med Image Anal Jan

2012;16(1):216–26. http://dx.doi.org/10.1016/j. media.2011.07.004 [Epub 2011 Jul 23]. Quellec G, Lamard M, Cochener B, Decencière E, Lay B, Chabouis A, et al. Multimedia data mining for automatic diabetic retinopathy screening. Conf Proc IEEE Eng Med Biol Soc 2013;2013:7144–7. http://dx.doi.org/10.1109/ EMBC.2013.6611205.

106. Prentasic P, Loncaric S. Weighted ensemble based automatic detection of exudates in fundus photographs. Conf Proc IEEE Eng Med Biol Soc 2014;2014:138–41. http:// dx.doi.org/10.1109/ EMBC.2014.6943548.
107. Zhang B, Vijaya Kumar BVK, Zhang D. Detecting diabetes mellitus and nonproliferative diabetic retinopathy using tongue color, texture, and geometry features. IEEE Trans Biomed Eng 2014;61(2):491–501.
108. Ogunyemi O, Kermah D. Machine learning approaches for detecting diabetic retinopathy from clinical and public health records. AMIA Annu Symp Proc Nov 5 2015;2015:983–90 [eCollection 2015].
109. Torok Z, Peto T, Csosz E, Tukacs E, Molnar A, Maros-Szabo Z, et al. Tear fluid proteomics multimarkers for diabetic retinopathy screening. BMC Ophthalmol Aug 7 2013;13(1):40. http://dx.doi. org/10.1186/1471-2415-13-40.
110. Jelinek HF, Wilding C, Tinley P. An innovative multi-disciplinary diabetes complications screening programme in a rural community: a description and preliminary results of the screening. Aust J Prim Health 2006;12:14–20.
111. Wright AP, Wright AT, McCoy AB, Sittig DF. The use of sequential pattern mining to predict next prescribed medications. J Biomed Inform Feb 2015;53:73–80. http:// dx.doi.org/10.1016/j. jbi.2014.09.003 [Epub 2014 Sep 16].
112. Deja R, Froelich W, Deja G. Differential sequential patterns supporting insulin therapy of new-onset type 1 diabetes. Biomed Eng Online Feb 21 2015;14:13. http://dx. doi.org/10.1186/s12938-015-0004-x.
113. Herrero P, Pesl P, Reddy M, Oliver N, Georgiou P, Toumazou C. Advanced insulin bolus advisor based on run-to-run control

and case-based reasoning. IEEE J Biomed Health Inform May 2015;19(3):1087–96.

114. Adem Karahoca M. Alper Tunga: dosage planning for type 2 diabetes mellitus patients using indexing HDMR. Expert Syst Appl 2012;39(8):7207–15.

115. Namayanja J, Janeja VP. An assessment of patient behavior over time-periods: a case study of managing type 2 diabetes through blood glucose readings and insulin doses. J Med Syst Nov 2012;36(Suppl. 1):S65–80. http://dx.doi.org/10.1007/ s10916–012–9894-3 [Epub 2012 Oct 27].

116. Shoombuatong W, Prachayasittikul V, Anuwongcharoen N, Songtawee N, Monnor T, Prachayasittikul S, et al. Navigating the chemical space of dipeptidyl peptidase- 4 inhibitors. Drug Des Devel Ther Aug 10 2015;9:4515–49. http://dx.doi.org/10. 2147/ DDDT.S86529 [eCollection 2015].

117. Patra JC, Chua BH. Artificial neural network-based drug design for diabetes mellitus using flavonoids. J Comput Chem 2011;32(4):555–67.

118. Schrom JR, Caraballo PJ, Castro MR, Simon GJ. Quantifying the effect of statin use in pre-diabetic phenotypes discovered through association rule mining. AMIA Annu Symp Proc Nov 16 2013;2013:1249–57 [eCollection 2013].

119. Bujac S, Del Parigi A, Sugg J, Grandy S, Liptrot T, Karpefors M, et al. Patient characteristics are not associated with clinically important differential response to dapagliflozin: a staged analysis of phase 3 data. Diabetes Ther Dec 2014;5(2): 471–82. http://dx.doi.org/10.1007/s13300–014-0090-y [Epub 2014 Dec 12].

120. Liu H, Xie G, Mei J, Shen W, Sun W, Li X. An efficacy driven approach for medication recommendation in type 2 diabetes treatment using data mining techniques. Stud Health Technol Inform 2013;192:1071.

121. Lee YC, Lee WJ, Liew PL. Predictors of remission of type 2 diabetes mellitus in obese patients after gastrointestinal surgery. Obes Res Clin Pract Dec 2013;7(6): e494–500. http://dx.doi.org/10.1016/j.orcp.2012.08.190.

122. Lee WJ, Chong K, Chen JC, Ser KH, Lee YC, Tsou JJ, et al. Predictors of diabetes remission after bariatric surgery in Asia. Asian J Surg Apr 2012;35(2):67–73. http://dx. doi.org/10.1016/j.asjsur.2012.04.010 [Epub 2012 May 26].

123. Zeevi D, Korem T, Zmora N, Israeli D, Rothschild D, Weinberger A, et al. Personalized nutrition by prediction of glycemic responses. Cell Nov 19 2015;163(5): 1079–94. http://dx.doi.org/10.1016/j.cell.2015.11.001.

124. Kaprio J, Tuomilehto J, Koskenvuo M, et al. Concordance for type 1 (insulin-dependent) and type 2 (non-insulin-dependent) diabetes mellitus in a population-based cohort of twins in Finland. Diabetologia 1992;35:1060–7.

125. Anjos S, Polychronakos C. Mechanisms of genetic susceptibility to type 1 diabetes: beyond HLA. Mol Genet Metab 2004;81:187–95.

126. Zhao LP, Bolouri H, Zhao M, Geraghty DE, Lernmark Å, Better Diabetes Diagnosis Study Group. An object-oriented regression for building disease predictive models with multiallelic HLA genes. Genet Epidemiol May 2016;40(4):315–32. http://dx. doi.org/10.1002/gepi.21968.

127. Nguyen C, Varney MD, Harrison LC, Morahan G. Definition of high-risk type 1 diabetes HLA-DR and HLA-DQ types using only three single nucleotide polymorphisms. Diabetes Jun 2013;62(6):2135–40. http://dx.doi.org/10.2337/db12-1398 [Epub 2013 Feb 1].

128. Park SH, Lee JY, Kim S. A methodology for multivariate phenotype-based genomewide association studies to mine pleiotropic genes. BMC Syst Biol 2011;5 Suppl. 2: S13. http://dx.doi.org/10.1186/1752-0509-5-S2-S13 [Epub 2011 Dec 14].

129. Lopes M, Kutlu B, Miani M, Bang-Berthelsen CH, Størling J, Pociot F, et al. Temporal profiling of cytokine-induced genes in pancreatic β-cells by meta-analysis and network inference. Genomics Apr 2014;103(4):264–75. http://dx.doi.org/10. 1016/j.ygeno.2013.12.007 [Epub 2014 Jan 24].

130. Lee J, Keam B, Jang EJ, Park MS, Lee JY, Kim DB, et al. Development of a predictive model for type 2 diabetes mellitus using genetic and clinical data. Osong Public Health Res Perspect

Sep 2011;2(2):75–82. http://dx.doi.org/10.1016/j.phrp.2011. 07.005 [Epub 2011 Aug 4].

131. Yarimizu M, Wei C, Komiyama Y, Ueki K, Nakamura S, Sumikoshi K, et al. Tyrosine kinase ligand-receptor pair prediction by using support vector machine. Adv Bioinforma 2015; 2015:528097. http://dx.doi.org/10.1155/2015/528097 [Epub 2015 Aug 11].

132. Global burden of diabetes. International Diabetes federation. Diabetic Atlas Fifth Edition 2011, Brussels. Available at http://www.idf.org/diabetesatlas. [Accessed 18th December 2011].

133. Pakhomov SV, Shah ND, Van Houten HK, Hanson PL, Smith SA. The role of the electronic medical record in the assessment of health related quality of life. AMIA Annu Symp Proc 2011; 2011:1080–8 [Epub 2011 Oct 22].

134. Nimmagadda SL, Dreher HV. On robust methodologies for managing public health care systems. Int J Environ Res Public Health Jan 17 2014;11(1):1106–40. http://dx. doi.org/10.3390/ijerph110101106.

135. Renard LM, Bocquet V, Vidal-Trecan G, Lair M-L, Couffignal S, Blum-Boisgard C. An algorithm to identify patients with treated type 2 diabetes using medicoadministrative data. BMC Med Inform Decis Mak 2011; 11:23.

136. Bradley PS. Implications of big data analytics on population health management. Big Data Sep 2013;1(3):152–9. http://dx.doi.org/10.1089/big.2013.0019 [Epub 2013 Sep 5].

137. Lee J, Giraud-Carrier C. Results on mining NHANES data: a case study in evidencebased medicine. Comput Biol Med Jun 2013;43(5):493–503. http://dx.doi.org/10. 1016/j.compbiomed.2013.02.018 [Epub 2013 Mar 18].

138. Tapak L, Mahjub H, Hamidi O, Poorolajal J. Real-data comparison of data mining methods in prediction of diabetes in Iran. Healthc Inform Res Sep 2013;19(3): 177–85. http://dx.doi.org/10.4258/hir.2013.19.3.177 [Epub 2013 Sep 30].

6 PROGRAMMABLE GENOME EDITING TOOLS AND THEIR REGULATION FOR EFFICIENT GENOME ENGINEERING

Tuhin Kumar Guha, Alvan Wai, Georg Hausner

Department of Microbiology, University of Manitoba, Winnipeg, Manitoba R3T2N2, Canada

ABSTRACT

Targeted genome editing has become a powerful genetic tool for studying gene function or for modifying genomes by correcting defective genes or introducing genes. A variety of reagents have been developed in recent years that can generate targeted double-stranded DNA cuts which can be repaired by the error-prone, non-homologous end joining repair system or via the homologous recombination-based double-strand break repair pathway provided a suitable template is available. These genome editing reagents require components for recognizing a specific DNA target site and for DNA-cleavage that generates the double-stranded break. In order to reduce potential toxic effects of genome editing reagents, it might be desirable to control the in vitro or in vivo activity of these reagents by incorporating regulatory switches that can reduce off-target activities and/or allow for these reagents to be turned on or off. This review will outline the various genome editing tools that are currently available and describe the strategies that have so far been employed for regulating these editing reagents. In addition, this review will examine potential regulatory switches/strategies that can be employed in the future in order to provide temporal control for these reagents.

Keywords: Meganuclease, TALEN, Zinc finger nuclease, CRISPR/ Cas9, Regulatory switch, Hammerhead ribozyme

INTRODUCTION

One of the challenges in biotechnology has been developing efficient and reliable ways to make targeted changes within the genome of cells. Traditional approaches of mutagenesis utilizing chemical agents or transposons can require extensive screening in order to recover desired mutations [1–6]. Genome editing strategies using double-stranded (ds) DNA viral vectors in differentiated human cells and RNA interference (RNAi) mediated targeted gene knockdown approaches also have some pitfalls [7–10]. For example, the protein composition of the viral capsid can be potentially immunogenic. Moreover, abnormal gene expression along with insertional mutagenesis may be triggered if there are random mutations in the viral sequences. On the other hand, the use of exogenously introduced dsRNA in RNAi technology can disrupt the "homeostasis" of the cellular machinery involved in gene silencing. Currently, the most popular genome engineering techniques apply DNA-cutting enzymes/complexes that generate targeted double-strand cuts [11–13], which are repaired by the host cells by either the error-prone, non-homologous end joining repair system (NHEJ), or the homologous recombination-based double-strand break repair pathway (HDR) [14–18]. The most frequent application of these endonuclease-based tools is the study of gene function through the inactivation of the target gene [19–21]. In addition, by providing a repair template, these systems allow for gene replacement strategies by taking advantage of the host cell's dsDNA break homologous repair system [22–24]. These new methods have tremendous potential towards the development of more accurate cellular and humanized laboratory animal models for various pathological conditions [25,26]. Moreover, these endonuclease-based genetic engineering techniques are being developed as therapeutic agents to cure human monogenic diseases [27–31]. Genome editing tools have far-reaching implications in the agricultural sector and in their potential of curbing pest populations, such as malaria insect vectors, or invasive species, such as cane toads and carps [32–36]. The latter applications are achieved in promoting the 'gene drive' of an introduced genetic element (such as a meganuclease) within an interbreeding population that can distort sex ratios (daughterless generations), or target genes related to fertility or pathogenicity [37–41].

Genome editing tools include meganucleases (MNs) [42–45], zinc finger nucleases (ZFNs) [46–49], transcription activator-like effector nucleases (TALENs) [50–53], clustered regularly interspaced short palindromic repeat (CRISPR)-associated nuclease Cas9 [54–56], and targetrons [57–63]. All of them can achieve precise genetic modifications by inducing targeted DNA double-strand breaks (DSBs). Depending on the cell cycle stage, as well as the presence or absence of a repair template with homologous terminal regions, the DSB may then be repaired by either NHEJ or HDR [64–68]. NHEJ can result in frameshift mutations that usually lead to gene disruption or gene knockout and/or the production of nonfunctional truncated proteins [69–71]; one exception being when a frameshift mutation was introduced to correct a defective coding sequence in the dystrophin gene [72,73]. In contrast, when single- or double-stranded DNA templates with homologous sequences that correspond to sequences flanking the break site are introduced within the cell, the lesion may be repaired using the HDR machinery [74,75].

One crucial concern when applying these genetic editing tools is the potential of cleavage at non-targeted sites. This event can be lethal or generate undesirable mutations resulting in the requirement of extensive screening in order to identify cells with the desired site-specific modifications. Many excellent reviews are available with regards to the above listed genome editing tools [13,21,42,44,45,76–87]. Therefore, this review will provide only a brief overview of the current genome editing tools and note any modifications made within recent years. The major focus in this review is to examine the efforts that have been made in the development of programmable, endonuclease-based platforms and various molecular switches that could be employed for the temporal regulation of these DNA-cutting enzymes in order to reduce off-target activities. The term "programmable" refers to the ability to engineer the nuclease-based platforms for recognizing various target sites (i.e. target specificity) in the genome.

GENOME EDITING REAGENTS

In general, genome editing tools using DSB nuclease-driven reactions (Fig. 1) can be divided into two groups. The first group consists of MNs, ZFNs and TALENs, which achieve sequence-specific DNA-binding via protein-DNA interactions [13,42]. The second group is comprised of two sub-groups: (i) CRISPR/Cas9 and targetrons, which are RNA-guided systems [56,57] and (ii) peptide nucleic acids (PNAs), triplex-forming oligonucleotides (TFOs), and structure-guided endonucleases (SGNs), which are DNA-based-guided systems [88–92]. A generalized comparison for the more commonly used genome engineering tools is presented in Table 1.

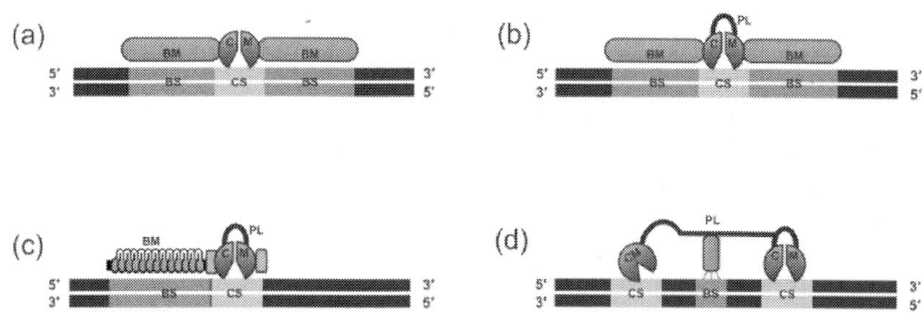

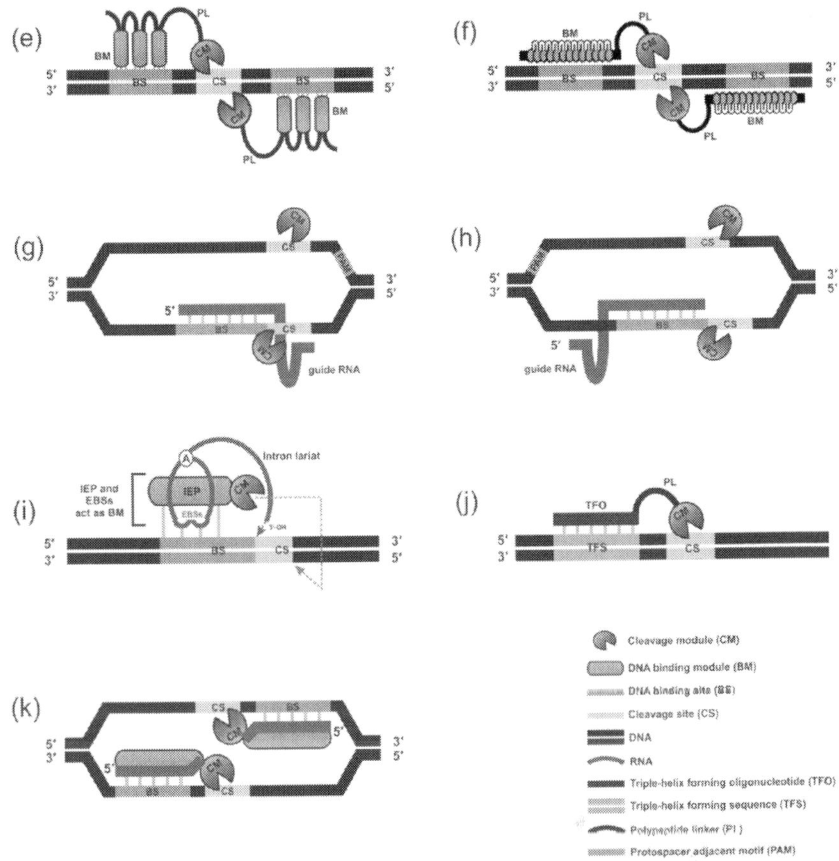

Figure 1: Examples of programmable genome editing tools. (a) Single-motif LAGLIDADG homing endonucleases, (b) double-motif LAGLIDADG homing endonucleases, (c) megaTAL, (d) MegaTev, (e) zinc-finger nucleases (ZFN), (f) transcription activator-like effector nucleases (TALENs), clustered regularly interspaced short palindromic repeats (CRISPR)/CRISPR-associated proteins (Cas) systems using (g) Cas9 or (h) Cpf1, (i) targetrons, (j) triplex-forming oligonucleotide (TFO) nucleases, and (k) structure-guided nucleases (SGNs). EBS = exon-binding site; IEP = intron-encoded protein. The nuclease domain of FokI is used to engineer ZNFs, TALENs, and SGNs. Elements of this figure have been adapted from Hafez et al. [44] NRC Research Press License number: 3981970186164.

Table 1: Generalized comparison of various genome engineering tools.

Nuclease platform	**MN**	**ZFN**	**TALEN**	**Targetron**	**CRISPR/Cas**
Source	Organellar DNA, Bacteria, Phages	Bacteria, Eukaryotes	Bacteria (*Xanthamonas*sp.)	Organellar DNA, Bacteria, Phages	Bacteria (*Streptococcus*sp.)a
Number of component(s)	1	2	2	2	1–2 (depends)b
Availability of core componentsc	Restricted	Available	Available	Restricted	Available
Type of recognition	Protein-DNA	Protein-DNA	Protein-DNA	RNA-DNA	RNA-DNA
Recognition site (bp)	18–44d	18–36	24–40	14–15	17–23
Double strand break pattern	Staggered cut (4 nt, 3′ overhang)	Staggered cut (4–5 nt, 5′ overhang)	Staggered cut (Heterogeneous overhangs)	Staggered cute	SpCas9 creates blunt ends; Cpf1 creates staggered cut (5′ overhang)
Function	Nuclease, Nickase	Nuclease, Nickase	Nuclease, Nickase	Site-specific bacterial gene disruptionf	Nuclease, Nickase
Best suited for	Gene editing	Gene knockout, Transcriptional regulation	Gene knockout, Transcriptional regulation	Gene knockout	Gene knockout, Transcriptional regulation, Base editing
Ease of design	Difficult	Difficult; Design of new ZFNs is much easier than MNs	Moderate	Moderate	Easy
Dimerization required	No	Yes	Yes	No	No
Ease of generating large scale libraries	Laborious	Laborious	Moderately laborious	Unknown	Easy
Specificity	High	Low–Moderate	Moderate	Moderate	Low–Moderateg

Multiplexing	Low	Low	Moderately high	Low	High
Gene drive	Possible	Unknown	Unknown	Unknown	Possible
Improved/other versions	MegaTEV, MegaTAL	AZP-SNase	Tev-mTALEN	Thermotargetron	Cpf1, eSpCas9
Cost (USD)h	4000–5000	5–10,000	< 1000	450–1500	< 100
Targeting constraints	Chromatin compaction	Non-guanosine rich sequence hard to target	5′ targeted base must be thymine for each TALEN monomer	Entry of RNP complex in nucleus difficult	PAM sequence must follow target site
Efficiency/Inefficiency	Small size of MN allows use in a variety of viral vectors	Small size of ZFN expression cassettes allows use in a variety of viral vectors	Large size of each TALEN makes it difficult to pack in viral vectors	Large size of ribonucleoprotein complex makes it difficult for entry into nucleus	Commonly used Cas9 from *S. pyogenes* is large, impose packaging problems in viral vectors i
Methylation sensitive	Yes	Yes	Yes	Unknown	No
First use in human cells	1994	2003	2011	2015	2013
Immunogenicity	Unknown	Low	Unknown	Unknown	Unknown
Vector packagingj	Multiple	Multiple	Few	Multiple	Multiple
Size of mRNA transcripts	Short	Short	Long	Short	Long
Mode of ex vivo delivery in animal cells	Electroporation, Viral transduction, Direct injection into zygotes	Electroporation, Lipofection, Viral transduction, Direct injection into zygotes	Electroporation, Lipofection, Viral transduction, Direct injection into zygotes	Electroporation, Lipofection	Electroporation, Lipofection, Viral transduction, Direct injection into zygotes

Source [13,21,45], Number of component(s) [80], Availability of core components [80], Type of recognition [81], Recognition site (bp) [42,49,51,55,57], Double strand break pattern [42,79], Function [45,76–80], Best suited for [13,45,162], Ease of design [77], Dimerization required [76], Ease of generating large scale libraries [77], Specificity [86], Multiplexing [77], Gene drive [37–41], Improved/ other versions [59,103,108,117,124,155,161], Cost (USD) [86], Targeting constraints [77], Efficiency/Inefficiency [77], Methylation sensitive [76,101], First use in human cells [80], Immunogenicity [77], Vector packaging [86], Size of mRNA transcripts [80], Mode of ex vivo delivery in animal cells [77,87].

A Most widely used Cas9 is from *Streptococcus pyogenes*. However, Cas9 orthologs, such as the smaller Cas9 proteins from *Streptococcus thermophilus* CRISPR1 (ST1), *N. meningitidis* (NM) and the large Cas9 protein from *Treponema denticola* (TD), have shown promising results in genome editing [154].

B 1 (if using a complex guide RNA with Cas9 protein) or 2 (if guide RNA and Cas9 delivered separately) [77].

C Availability of core components refers whether the building blocks are restricted to industry, available through and academic collaboration/ purchase, or readily and freely available from not for profit agencies or commercial DNA synthesis [86].

D The range used here, encompasses a number of different MNs and not only LAGs. The largest recognition site of a LAG is ~ 31 bp [94].

E The 3′ hydroxyl group of the group II intron serves as a nucleophile and cleaves just one strand of the DNA homing site. The RNA lariat is reverse spliced into the target site and the endonuclease domain of the assisted protein partner cleaves the complementary DNA strand [57].

F Although compromised activity is observed in eukaryotes and mammalian system due to the suboptimal codon usage, translational repression of the RT, nonsense-mediated decay (NMD) of group II intron-containing RNAs and suboptimal magnesium ion (Mg^{+2}) concentrations, this RNA-guided endonuclease (RGEN) has shown potential for high site-specific retargeting in prokaryotes by reprogramming the intron EBS [194].

G Recently improved specificity has been reported for eSpCas9 enzyme [161].

H Approximate cost required to generate a single, gene specific candidate reagent [86].

I Short oligonucleotides may be packaged along with guide RNAs into a single adeno-associated virus [154].

J Vector packaging refers to the reagents ability to be packaged and delivered in multiple delivery vehicles. However, the size of TALENs makes them the most restrictive in this regard. To date, only one version (derived from *S. aureus*) of CRISPR/Cas9 can be packaged in an adeno-associated viral vector [86].

Meganucleases, or homing endonucleases (HEases; Fig. 1a,b), are highly site-specific dsDNA endonucleases that can be reengineered to expand their target site repertoires using various strategies, such as computational structure-based design, domain swapping, combined with yeast surface display for efficient detection of HEases with desired sequence specificities [93–98]. The LAGLIDADG family of MNs have been extensively studied and applied as genome editing tools [43,44,45,99–101]. Unless otherwise mentioned, we are referring to LAGLIDADG enzymes as MNs for simplicity. One essential drawback for this class of enzyme is its non-modular configuration. The DNA recognition and cleavage functions can be, in part, intertwined in a single protein domain. Therefore, engineering of MNs has been challenging [45,76] and has resulted in the development of other editing tools. However, a recent study suggests that there are multiple points across the LAGLIDADG protein that can be involved in holding metal ions in suitable positions to facility cleavage [102]. This finding along with technologies, such as yeast surface display-SELEX, still hold promise for MNs to be engineered more efficiently in the near future [97]. Moreover, a single-chain modular nuclease architecture, termed 'megaTAL' (Fig. 1c), was designed in which the DNA-binding region of a transcription activator-like (TAL) effector is appended to a site-specific MN for cleaving a desired genomic target site [103]. The latter synthetic version of a MN provides a modular design, separating the endonuclease and DNA binding activities. Therapeutic applications that demand precision with regards to gene modification activity can be addressed by these engineered variants of MNs, as they are considered to be highly target-specific 'molecular scissors' [45]. MNs are also in demand as components of vector/cloning systems (e.g. HomeRun vector assembly system) and synthetic biology applications (e.g. iBrick) that require rare-cutting enzymes [104,105].

Even though the NHEJ pathway is usually exploited to introduce mutations at the DSBs within the genome [15,106], sometimes,

DSBs possess compatible "sticky" ends that can be repaired without any introduced mutation [107]. Recently, the 'MegaTev' (Fig. 1d) architecture has been generated which involves fusion of the DNA-binding and cutting domain from a meganuclease (Mega, I-OnuI) with another nuclease domain derived from the GIY-YIG HEase (Tev, I-TevI). This protein was designed to position the two cutting domains ~ 30 bp apart on the DNA substrate and generate two DSBs with non-compatible single-stranded overhangs for more efficient gene disruption [108]. More recently, similar to the MegaTev concept, Wolfs et al.have designed another dual nuclease, in which the Tev endonuclease domain is attached to the Cas9 nuclease domain, known as TevCas9 [109]. This hybrid nuclease, when introduced within human embryonic kidney cells (HEK293) along with appropriate guide RNAs, has been shown to delete 33 to 36 bp of the target site, thereby creating two non-compatible DNA breaks at moderately higher frequencies (40%). Therefore, this newly designed dual active endonuclease also promises to favor genome editing events (i.e. introduce mutations) by avoiding the creation of compatible "sticky" ends which lead to a failed attempt of genome editing [109].

More recently developed genome editing tools try to be more flexible with regards to retargeting the reagent to different sequences by having a modular design: a DNA-cutting domain (that can be non-specific) and a distinct programmable DNA-binding domain. The ZFNs are artificial endonucleases that have been generated by combining a small zinc finger (ZF; ~ 30 amino acids) DNA-binding/recognition domain (Cys_2His_2) to a type IIS nonspecific DNA-cleavage domain from the FokI restriction enzyme (Fig. 1e). However, the cleavage activity of the FokI endonuclease demands dimerization [46,110]. As a ZF module recognizes a 3 bp sequence, there is a requirement for multiple fingers in each ZFN monomer for recognizing and binding to longer DNA target sequences [46]. In the past, using structure-based design, two ZFN variants were engineered that efficiently cleaved DNA only when paired as a heterodimer, thereby providing a potential avenue for improving the specificity of ZFNs as gene modification reagents [111]. In a different structure-based study, using 3D protein modeling and energy calculations through computer-based softwares, researchers have identified potential residues within the FokI dimer interface that are responsible for ZFN dimerization [112]. These newly designed ZFNs were considered significantly less genotoxic (i.e. cleavage at on-target sites) in the cell-based recombination studies because the homodimerization could be prevented by lowering the dimerization energy, hence prevent activation of the dimeric FokI [112].

Recently, ZFNs have been used as a potent antiviral therapy in the inactivation of specific coreceptors, thereby protecting cells from the viral entry in order to establish infection [113]. Even though ZFNs showed impressive results in modifying the HIV CCR5 coreceptor surface protein in the autologous CD4 T lymphocytes of persons infected with HIV [114], there is still the risk of cleavage at ectopic sites due to the modular architecture of ZFNs and the non-specific nature of FokI [49,115].

Apart from implementing ZFNs as genome editing tools [48,49], recently, the artificial zinc-finger protein (AZP)-staphylococcal nuclease (SNase) hybrid was designed (AZP-SNase) for potential antiviral therapies. This artificial nuclease can bind and cleave a specific origin of replication sequence of the human papillomavirus type 18 (HPV-18) thereby inhibiting viral replication in mammalian cells [116]. However, one disadvantage of this reagent is that the SNase has been shown to cleave both single and double-stranded RNA as well as the host DNA (single or double-stranded). Further modification involving switching of the SNase moiety in the AZP-SNase to the single-chain FokI dimer (scFokI) cleaved the viral DNA. Therefore, this newly designed hybrid ZFN is expected to serve as a novel antiviral reagent for inactivating human DNA viruses with fewer side effects [117].

TALENs are artificial endonucleases (Fig. 1f) designed by fusing the DNA-binding domain (multiples of nearly identical repeats each comprised of ~ 34 amino acids) obtained from TAL (transcription activator-like) effector (TALE) protein to the cleavage domain of the FokI endonuclease [118]. Each TALE repeat independently recognizes its corresponding nucleotide (nt) base with two variable residues [termed the repeat variable di-residues (RVDs)] such that the repeats linearly represent the nucleotide sequence of the binding site. Despite the tolerance to mismatches of longer TALENs in vitro, they seem to have higher genome editing activity and considered less genotoxic than ZFNs [119–123]. TALENs can be redesigned to bind user-defined sequences by simply joining appropriate repeat units. Like ZFNs, TALENs are dimeric in nature; this necessitates the design of two independent DNA-binding modules to target a single sequence. One advantage of the requirement for dimerization is enhanced specificity over monomeric enzymes [50,51]. Although the FokI enzyme is useful in terms of flexibility in the choice of various target sites, its nonspecific activity also increases the probability for more frequent cleavage at off-target sites in the genome [124]. As an alternative approach to the FokI-based architecture, monomeric Tev-TALE nucleases (Tev-mTALENs) were created. Here, the sequence-specific, monomeric nuclease domain from the I-TevI HEase is fused with TALEs. Thus, only a single DNA-binding module is needed to target a sequence for cleavage. However, the use of a domain with predetermined recognition requirements, like TevI, significantly limits the range of genomic targets [124].

Components derived from the bacterial "immunity" system, CRISPR locus and the Cas9 nonspecific endonuclease (CRISPR/Cas9), form a novel RNA-guided endonuclease (RGEN; Fig. 1g) for precise and efficient gene targeting [125–128]. The uniqueness of this platform is based simply on designing guide RNAs (gRNAs) essentially serving as CRISPR RNAs (crRNAs) that are bound by the Cas9 nuclease. Initially, the gRNAs were expressed separately as trans-activating CRISPR RNA (tracrRNA) and the "user-designed" crRNA sequence, both of which are chemically synthesized for the effective targeting and cleavage of a sequence within the gene of interest [129]. More commonly, for simplicity, both the crRNA and tracrRNA are expressed as a single construct known as single guide RNA (sgRNA) [55]. Cas9, however, does not require any engineering for retargeting. Complementary base pairing allows a segment of the gRNA sequence (~ 18–20 nt) to hybridize with the targeted DNA sequence and thus docking of the Cas9 nuclease at that location. The H–N–H and the RuvC nuclease domain of the Cas9 cleave both DNA strands to create DSBs 3 bp upstream (5′) of the protospacer adjacent motif (PAM) sequence. The PAM sequence is specific to each Cas9 nuclease obtained from different bacterial species [130,131]. Therefore, different sources for Cas9 have to be explored with regards to optimizing this system to a wide range of eukaryotes/target sequences. Eventually, by designing various gRNAs, this system can be utilized for targeted mutagenesis by inducing the NHEJ pathway or it can be applied to repair or replace alleles by utilizing the cellular HDR repair mechanism with the presence of a user-provided DNA corrective template.

A modified version of the RNA-guided Cas9 has been developed that allows for "targeting" regulatory sequences and manipulating gene expression. For this purpose, a nuclease-deficient version of Cas9 protein has been generated by mutating positions H840A in the H-N-H domain and D10A in the RuvC domain [132]. This variant is commonly known as "dead" Cas9 or dCas9. However, the DNA-binding characteristic remains unaffected for this modified protein [55,133]. Therefore, gene silencing (referred to as CRISPR interference or CRISPRi) or gene activation can be made possible by fusing dCas9 with various effector domains [134–140].

In a recent study, GCaMP (a calcium-sensitive modified GFP) fluorescence signals were monitored in induced pluripotent stem cells (iPSCs) to determine if CRISPRi, based on the RNA-guided dCas9 being targeted to bind to a specific promoter sequence, can knock down GCaMP expression and whether removal of doxycycline [tetracycline (Tet) derivative] from the culture reversed its expression. Expression of the CRISPRi components are under the control of the Tet-response element (TRE), thus doxycycline acts as an inducer for the regulatory protein that interacts with the TRE. The researchers found that GCaMP expression was downregulated by 98% after addition of doxycycline for 7 days. However, the expression was completely restored after removing doxycycline for 14 days [141]. This proof of principle study demonstrated that reversible RNA interference is possible with regulated versions of dCas9 and this might become a powerful alternative to RNAi, which can be applied to knock down expression of a gene but cannot be reversed. Furthermore, dCas9 has been repurposed as a visualization tool. For example, Enhanced Green Fluorescent Protein (EGFP), when fused with dCas9, enabled visualization of both repetitive and nonrepetitive DNA sequences [142]. Recently, the dCas9 has also been used as a building block for RNA-guided FokI nucleases, thereby dCas9 also has applications in genome editing. Here, the dCas9 and its sgRNA has been recruited as a DNA-binding module that is coupled with FokI, which serves as the nuclease component [143]. This reagent requires dimerization that is brought about by the FokI-dCas9 fusion proteins being recruited to sequences adjacent to the target site by two different gRNAs.

Multiplex editing is possible with CRISPR/Cas9 [144–148] and the PAM requirements of Cas9 do not place much of a limitation on target choice because PAMs are quite short sequences [125,131]. However, the risk of off-target activities exists [149,150]. Henceforth, paired Cas9 nickases and gRNA modifications, like truncated gRNA (tru-gRNA), have been constructed and have shown promising results with regards to reducing off-target activities [151–154].

Another recent innovation is the isolation of the novel CRISPR protein, Cpf1, a non-Cas9 CRISPR nuclease (Fig. 1h). Cpf1 has been shown to generate staggered double-strand breaks with "sticky ends" at targeted sites, which is not the case for Cas9 proteins [155]. The generation of sticky ends and the programmability of the CRISPR/Cpf1 endonuclease system make this reagent very suitable for developing DNA assembly strategies (e.g. C-Brick) [156]. Cpf1 requires a T-rich PAM sequence, making this reagent suitable for targeting T-rich segments within genomes [155,157]. Moreover, Cpf1 seems to have inherently higher specificity than currently available forms of Cas9 [158,159]. A variant of Cas9, recently described from *Staphylococcus aureus*, is considerably smaller (by 1 kb) compared to other bacterial Cas9 proteins. This represents an improvement as it allows for the design of more compact vector systems that are more easily accommodated within the more efficient viral-based delivery systems for in vivo or ex vivo applications [160].

Development of the "enhanced specificity" SpCas9 (eSpCas9) through structure-guided protein engineering has shown a dramatic decrease in off-target indel (insertions-deletions) formation, thereby contributed towards a significant improvement over the *Streptococcus pyogenes* Cas9 (SpCas9) enzyme [161]. In this study several SpCas9 mutants were designed by substituting 32 positively-charged residues, which are responsible for recognizing the nucleotide groove, with individual alanine moiety. Then after, using a previously validated guide sequence, these single amino acid SpCas9 mutants were tested for specificity by targeting them to the *EMX1* target site in human embryonic kidney (HEK) cells. With these improved versions, the specificity of indel formation at the target sites has been shown to be improved by a factor of 2 to 5 [161].

Usually, genome editing tools introduce dsDNA breaks at a target locus. However, a recent study has shown that one could bypass the need for dsDNA backbone cleavage and the required introduction of a donor template for genome editing. Strategies are being developed that harness enzymes that can edit DNA sequences by chemically modifying nucleotide bases [162]. For example, it has been shown that fusing rAPOBEC1 cytidine deaminase [163], which showed the highest deaminase activity among the four different deaminase enzymes tested, to the amino-terminus of dCas9, does not affect the deaminase activity. Therefore, 'base editing' using cytidine deaminase may be an alternative new approach to genome editing that enables irreversible conversion of one target DNA base into another. In that study, direct conversion of cytidine to uridine in a programmable manner has been shown to be possible with the help of a guide RNA [162]. However, how would uridine, which is one of the building blocks of RNA, be tolerated within the DNA sequence is questionable. Usually, cytidine deaminases use RNA as the substrate and, interestingly, a few of them have been reported to work on single stranded (ss) DNA. Fortuitously, when dCas9-target DNA complex is formed, the displaced DNA strands are separated to form the 'R-loop' complex whereby both the strands are separated. This conformation might serve as an efficient substrate for this programmable conversion of cytidine to uridine in DNA. One of the major challenges of this technique is that it is unable to perform precise base editing, in particular when multiple cytidines are present in close proximity, i.e. the spreading of base modification to neighboring cytidine occurs [162].

Although not a form of genome editing, another noteworthy development is the nuclease-inactive *S. pyogenes* RNA targeting CRISPR/Cas9 (RCas9) protein that is conjugated with the green fluorescent protein [164]. This reagent has been engineered to bind to RNA with the aid of a sgRNA strands. The sgRNA allows for the system to be programmable, thereby allowing for endogenous RNA tracking in living cells [164].

The targetron (Fig. 1i) is a ribonucleoprotein particle (RNP) that consists of an engineered group II intron RNA lariat molecule and a multidomain group II intron-encoded protein [i.e. reverse transcriptase (RT)] which has been used for mutagenesis of bacterial genes [57–60]. The strategy is based on group II retrohoming where the intron lariat recognizes its native DNA target site by the presence of an "exon-binding sequence" (EBS) that can base pair with a corresponding "intron-binding" sequence (IBS) present within the targeted gene/site. These "EBS/IBS" interactions require homology for about ~ 14 bp [61]. This RNA-guided endonuclease system has shown potential for highly site-specific retro-targeting (mutagenesis by insertional mutations) of genes in prokaryotes by simply reprogramming the intron EBS to match target sequences within targeted genes [62]. Compromised activity is observed in eukaryotes, such as mammalian systems, due to suboptimal codon usage, translational repression of the RT, nonsense-mediated decay (NMD) of group II intron-containing RNAs, and suboptimal magnesium ion (Mg^{+2}) concentrations [78]. In addition, the entry of the targeting RNA, in the form of an RNP, into the nucleus or chromatin still remains the major obstacle for applications of targetrons among eukaryotes [63].

Synthetic molecules such as peptide nucleic acid (PNA) oligomers [88] and triplex-forming oligonucleotide (TFO; Fig. 1j) [165] have been developed as potential alternatives to the above outlined genome editing reagents. The strategy is to develop programmable DNA-binding modules that can be coupled to DNA-cutting domains. Although their use, so far, has been limited, they do offer some advantages that are worth mentioning. For example, PNAs have higher binding strength compared to oligonucleotides [166]. Therefore, designing long PNA oligomers for use in DNA-binding is not a prerequisite. This is in contrast with the targetron and the CRISPR/Cas systems, which usually require DNA-binding modules of 14–22 bases for efficient recognition and DNA-binding [55,57]. Moreover, PNAs can tolerate a wide pH range and are not easily recognized by either nucleases or proteases [167]. Also, improvements regarding the delivery within the cytoplasm have been made when different cell-penetrating peptides were coupled to PNAs by covalent bonds [167]. The TFO nucleases are sequence-specific type II restriction enzyme-TFO conjugates [165]. Instead of a protein-based DNA-binding domain, as seen in MNs, ZFNs, or TALENs, these DNA-binding oligonucleotides can be engineered to cater to various DNA target sites. However, the DNA-cutting components of TFO nucleases are activated by Mg^{+2} ions, and thus cleavage activity might be triggered before the RE-TFO conjugate assembles on the intended target site [168,169]. There are also versions of TFOs that operate as dimers and utilize FokI as the nuclease domain [170]. Interestingly, both PNAs and TFOs can be also designed to target RNA duplexes forming RNA triplexes, which may have potential application in gene regulation [171].

Another new entry among potential genome editing tools is the structure-guided endonuclease (SGN; Fig. 1k), which is composed of the flap endonuclease-1 (FEN-1) attached to the FokI nuclease domain [90]. In eukaryotes, FEN-1 is involved in DNA repair and DNA replication that involves the removal of 5′ overhanging flaps and in processing the 5′ ends of Okazaki fragments in lagging strand DNA synthesis [90,91]. The engineered SGN complex operates as a dimer and is guided to a target site by two single-stranded guide-DNAs (gDNAs, 20 to 60 nts). The gDNAs are designed to have a single-base mismatch at the 3′ end; i.e. a 3′ "flap" structure forms once these oligonucleotides have bound to their targets. The FEN-1 component of the SGN recognizes a 3′ "flap" structure and is recruited to this position. Thereafter, the Fok1 dimer will form and cleave the target DNA strands. This approach has been successfully demonstrated in zebrafish embryos and therefore has potential for genome editing among the metazoans [92]. It was noted that SGN can generate large deletions at the cut site, probably due to the combined activities of the FEN-1 and FokI nuclease domains. This might be an advantage when the goal is to achieve gene disruptions [92].

The CRISPR/Cas system has definitely expedited biological research with regards to genome editing. However, recent work involving the Argonaute family of proteins from *Natronobacterium gregoryi* hints at the possibility of another option for genome editing in mammalian cells [172]. Gao et al. (2016) noted that NgAgo (*N. gregoryi* Argonaute) with the aid of DNA oligonucleotides can be programmed for site-specific targeting. The 5′ phosphorylated single-stranded guide DNA (gDNA) is usually 24 nt long sequence, and when bound to the NgAgo protein it is sufficient to create a DSB at the corresponding DNA target site. This system has the potential to edit GC-rich regions within the genome and does not have a PAM sequence requirement, thus allowing for a wider range of genomic targets [172]. However, this work is currently under scrutiny as other groups noted that the work was not reproducible in their laboratories [173]. Therefore, considerable efforts may yet be required to demonstrate the promised utility of the NgAgo-gDNA based system for genome editing.

In addition to the above genome editing reagents, site-specific recombinases have been shown to work efficiently as genome engineering tools in mammalian cells [174,175]. These recombinases have been mostly derived from the bacteriophages, such as the Cre resolvase from the P1 phage of *Escherichiacoli* and phiC31 integrase from a phage of *Streptomyces* sp. [176,177]. These recombinases are highly site-specific and recognize long DNA binding sites of 34 bp. Unlike the above genome editing tools, these enzymes can process DNA strand exchange in a "cut and paste" fashion without creating any free DSB. This means that the complete recombination happens immediately in a concerted manner within the "all-in-one" recombinase enzyme complex, without being assisted by other cellular enzymes [178]. Typically, the phiC31 integrase assists in a unidirectional recombination between two different attachment (*att*) sites (*attB* and *attP*), resulting in the integration of a plasmid or any other DNA fragment quite precisely within the chromosome [179]. Fortuitously, along with *att* sites, the human genome and other larger genomes contain pseudo-*attP* sites [180]. With regards to human gene therapy applications it was noted that a variant of the phiC31 integrase (a 613-amino acid protein) can recognize these pseudo-*attP*sites, and thereby is able to insert DNA molecules, such as therapeutic genes or plasmids at preferred sequences within the mammalian genomes [181].

CURRENT REGULATABLE DNA-CUTTING ENZYMES

In some instances, such as in vivo or ex vivo gene targeting, temporal regulation of endonuclease activity might be desirable in order to minimize nonspecific activity of the DNA-cutting enzymes (Fig. 2). DNA-cutting enzymes ultimately can have mutagenic and/or toxic side effects if they go off-target. Previously, a reversible redox switch was developed that controlled the endonuclease activity of PI-SceI in vitro [182]. Here two cysteine amino acid residue pairs were inserted into the HEase DNA-binding loops to allow for disulfide bond formation (oxidizing condition) that locks the endonuclease into a nonproductive conformation. This can be reversed by reducing conditions that result in the breakage of the disulfide bond, thereby yielding an active conformation of the protein. Since the inside of cells have reducing environments, this approach is not practical for activating the enzyme during in vivo applications [182].

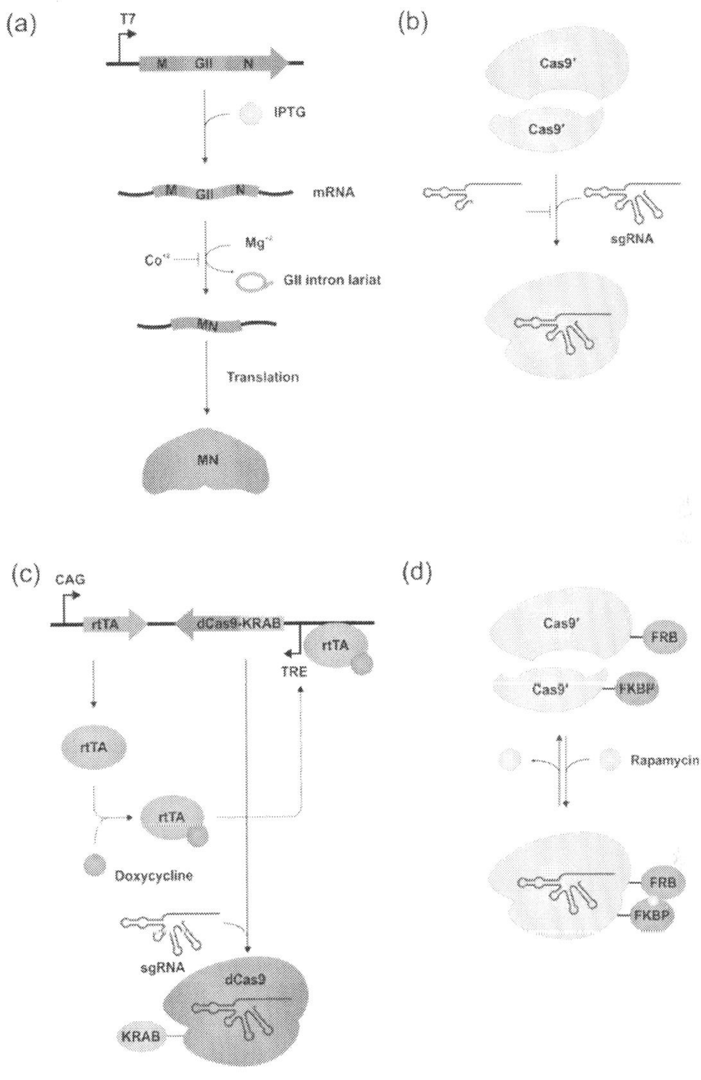

Figure 2: Strategies used to modulate Cas9 activity. (a) Group II intron (GII)-based switch, (b) separating Cas9 into two peptides, termed split-Cas9, (c) Tetracycline-inducible and reversible expression system, and (d) ligand-dependent dimerization of split-Cas9. Note: the strategy illustrated in (a) is based on the original study conducted by Guha and Hausner [185] on modulating expression of a meganuclease, not Cas9. A similar case is observed in (c), where Mandegar et al. [141] modulated the expression of dCas9, not Cas9. In both cases, a similar approach might also be possible with Cas9.

(e) Light-dependent dimerization of split-Cas9, termed photoactivatable Cas9 (paCas9), (f) intein-Cas9, which are activated by splicing of a ligand-dependent intein, (g) and unstable destabilizing domain-Cas9 (DD-Cas9) fusions, which are degraded unless provided with the ligand, Shield1.

Abbreviations: CAG = cytomegalovirus early enhancer/chicken β-actin promoter; Cas9 = clustered regularly interspaced short palindromic repeat (CRISPR)-associated protein 9; Cas9′ = partial Cas9; dCas9 = dead Cas9; FKBP = FK506 binding protein; FRB = FKBP-rapamycin binding; IPTG = isopropyl β-D-1-thiogalactopyranoside; KRAB = Krüppel-associated box; MN = meganuclease; mRNA = messenger RNA; rtTA = reverse tetracycline-controlled transcriptional activator; sgRNA = single-guide RNA; TRE = tetracycline response element; T7 = T7 RNA polymerase promoter; 4-HT = 4-hydroxytamoxifen; DD = destabilizing domain; nMag = negative Magnet; pMag = positive Magnet; sgRNA = single-guide ribonucleic acid. See text for more details.

Recently, it was shown that expression of an active MN, I-CthI [183], can be controlled, or at least attenuated, by the splicing activity of autocatalytic group II intron sequences (Fig. 2a) [184,185]. The expression and activity of I-CthI HEase was modulated in *E. coli* by inserting ribozyme type introns (group IIA and IIB intron sequences), that lack open reading frames (ORFs), separately into the MN ORF, where splicing of these introns could be stimulated by the addition of 5–10 mM $Mg^{+2}$and antagonized by the addition of 10 μM cobalt ions (Co^{+2}) in the bacterial growth media. Group II intron sequences are readily available [186–188], deposited in various databases, and these sequences could be coopted as regulatory switches [184] and unlike previous attempts to control MN activity via in vitro redox switches [182], in vivo regulation of endonuclease activity utilizing group II introns is possible [184,185]. In the future, with regards to group II intron-based "switches", one could achieve even tighter control by utilizing trans-splicing group II introns. Trans-splicing group II introns (or fragmented group II introns) have been noted in organellar genomes but it is unknown if these types of introns can function in *E. coli*[189,190]. However, it has been shown that the Ll.LtrB group II intron (including a version where the ORF was deleted) from the Gram-positive bacterium *Lactococcus lactis* can splice in *trans* when fragmented at various locations throughout its structure [191]. Therefore, a MN ORF could be split and encoded by two compatible plasmids carrying different selectable markers and different promoters. One construct can bear the amino-terminal part of the HEase ORF plus the 5′ segment of a group II intron sequence and the other construct can carry the 3′ segment of group II intron sequence plus the carboxyl-terminal part of the MN ORF. Upon expression, these two RNAs can assemble via the intron segments into a tertiary structure that promotes trans-splicing of the intron sequences. Thus, the exons get ligated together to produce a functional MN transcript. Even though group II intron-based "molecular switches" have been shown to work quite efficiently in bacterial systems, they may only have limited applications in eukaryotes. Compromised activity of group II intron splicing and retrohoming in nuclear environments has been noted to be due to the suboptimal intracellular Mg^{+2}concentrations [78]. In addition, intron-containing transcripts are subjected to NMD and translational repression [63]. However, recent work by Lambowitz's group showed progress towards developing a group II intron expression system that can circumvent expression/splicing barriers. They have shown that retrohoming into chromosomal target sites in human cells at appreciable frequencies is possible when Mg^{+2} salts are added to the culture medium [192,193]. Through genetic selections and deep sequencing techniques, they also identified several group II intron RNA mutations in the catalytic core domain V (DV) that partially rescued retrohoming in Mg^{+2}-deficient *E. coli*[194] and in human cells at low Mg^{+2} concentrations [78]. Their findings have implications in terms of demonstrating the feasibility of selecting various group II intron variants that function more efficiently at low Mg^{+2} concentrations. Also, recent characterization of group II introns that are less dependent on Mg^{+2} may offer new impetus on the utilization of group II ribozyme-based switches in eukaryotic systems [194]. For now, one can foresee the application of group II intron sequences as agents that allow for inducible genome editing in cell types that are suited towards supporting the splicing of these elements.

It has been documented that constitutive expression of the Cas9 enzyme is one of the problems limiting the use of the CRISPR/Cas9 systems. Constitutive expression or high dosage of the Cas9 can lead to an increase in indel frequencies at off-target sites thereby initiating a DNA damage response [79,149]. However, another study showed that the Cas9 enzyme alone is quite well tolerated, particularly in mice. Therefore, viable mouse models expressing Cas9 constitutively do exist [145]. Apart from transient delivery of purified Cas9:sgRNA complex into cellular environments [195,196] and regulating expression through the use of inducible promoters [197,198], several methods have been developed with regards to addressing the regulation of this enzyme. Initial attempts to separate or split the Cas9 protein into two fragments have been successful. The Cas9 protein was separated in two polypeptides, one expressing the nuclease lobe and the other expressing the α-helical lobe of the enzyme (Fig. 2b) [199]. The two modules interacted and combined only in the presence of a sgRNA, thereby restoring the activity of a full-length Cas9. The enzymatic activity of the holoenzyme formed from two peptide components was shown to be no different from that of the native Cas9 and therefore remained effective for genome editing in human cells when full-length sgRNAs were used. However, shortening or modifying the sgRNAs, particularly removing the hairpins 1 and 2 from the 3′ end of the sgRNA structure rendered the protein modules in a separated, inactive conformation [199].

As an alternative to the above, there are versions of the Cas9 enzyme that are split into two components that can reconstitute into an active Cas9-gRNA complex by the addition of chemical signals, such as doxycycline and rapamycin (Fig. 2c,d) [200,201]. Reversibility of these systems can be achieved upon the withdrawal of these ligands. Even though inducible methods based on plasmid constructs that included various regulatory elements that can be modulated to determine the expression of various CRISPR components have been used for generating conditional gene knockouts and reducing off-target effects during genome modification, one important concern still lurks regarding the adversities of these chemicals (i.e. inductants, ligands). Also ligands required for components to assemble at the protein level also may be of concern with regards to side effects on the cells, and this may limit their in vivo or ex vivo applications. For example, inducing the dimerization domains with rapamycin can perturb the endogenous mammalian target of rapamycin complex 1 (mTOR1) pathway leading to undesirable biological effects [200,202]. However, the possibility of building an array of other inducible split-Cas9 enzymes that utilizes the same concept but depend on other chemical-sensing domains, such as abscisic acid or gibberellin-sensing domains may be an effective alternative in terms of toxicity. The utility of these domains towards induction, however, needs to be tested before they can be introduced in animal or plant cells.

Light can be controlled both temporally (microseconds) and spatially (microns) and is noninvasive to biological systems (Fig. 2e) [203–205]. Therefore, regulating the activity of DNA-cutting enzymes using light as a trigger may be an alternative to the above described approaches. Recently, this concept was applied to the genome editing of human cells by engineering a photoactivatable Cas9 (paCas9) that allows for optogenetic/light control of the CRISPR/Cas9 system [206]. Briefly, paCas9 consists of split-Cas9 fragments, each appended to photoinducible dimerization domains named "Magnets". Both positive (pMag) and negative (nMag) "Magnets" are light inducible dimerization proteins (~ 150 amino acids each), which heterodimerize in response to blue light irradiation [207] and thereby reconstituting an active Cas9 protein. When expressed in HEK293T cells, the paCas9 proved effective in inducing targeted genome sequence modifications through both NHEJ and HDR pathways. Conversely, the components dissociated and the genome editing activity has been shown to turn off by simply extinguishing the light source [206].

In the past, it was shown that the catalytic activity of the PvuII restriction endonuclease (REase) could be controlled by a photoswitch involving a derivative from a bifunctional azobenzene [203]. However, unlike "Magnets", which heterodimerize and activate the paCas9 protein under blue light, the azobenzene-derivative photoswitch deactivates the PvuII REase under blue light and activates it only under illumination by ultraviolet (UV) light (wavelength ~ 365 nm). This system can be turned into a reversible photoswitch as the *trans* isomeric form of azobenzene locks the enzyme in the inactive "off" state, while the *cis* form of azobenzene engages the enzyme into the active "on" state. One important advantage of the photoinducible system is that chemically cross-linked endonuclease in the inactive state can be activated using an external signal light source for DNA-cleavage activity at the specific target sites after being successfully transported into the nucleus of the cell using an appropriate delivery system, such as cationic amphiphilic lipids [208]. One potential concern is that near UV light might be damaging to DNA [209,210].

Conditional activation of the Cas9 enzyme has also been developed by placing a 4-hydroxytamoxifen (HT)-responsive intein sequence (37R3-2) within the Cas9 ORF, where the intein has been engineered to splice from the host protein when a cell-permeable small ligand (4-HT) is added to the media (Fig. 2f) [211]. In the same study, when the HEK293-GFP cells were treated with 4-HT for 12 h, intein-Cas9 variants in combination with the sgRNAs that target the well-studied *EMX*, *VEGF* and *CLTA* loci exhibited substantially improved specificity compared to that of wild-type Cas9. The presence of 4-HT in the cell culture media increased the on-target modification frequency of the intein-Cas9(S219) variant by 3.4- to 7.3-fold than what was observed and statistically calculated in the absence of 4-HT. However, this system suffers from the reversibility issue in a way that, when the intein splices out of the Cas9 protein, it could not be turned off because the intein cannot be inserted back within the Cas9 ORF [211].

Recently, in order to address the periodic modulation of the Cas9 function, a chemical-inducible CRISPR/Cas9 system was developed, where switching the activity of the Cas9 (or iCas in this case) to both 'on' and 'off' states were possible [212]. In that study, the authors have shown that a tight spatiotemporal control over the Cas9 protein (iCas9) could be achieved by fusing two hormone-binding domain of the estrogen receptor (ERT2) on each terminus of the Cas9 protein (i.e. $(ERT2)_2$–Cas9–$(ERT2)_2$). In this configuration, Cas9 cannot enter the nucleus of human cells, thereby preventing the access to the genomic DNA for editing purpose. However, the addition of the ligand 4-HT permits the translocation of the fusion protein into the nucleus [212].

This ligand-based Cas9 activation approach can be used in conjunction with other strategies that are dedicated to reduce off-target issues, such as using paired Cas9 nickases [126], truncated guide RNAs [153], or FokI-dCas9 fusions [213]. In this context, we find that the ligand-dependent intein-based regulation is somewhat analogous to the group II intron ribozyme-based molecular switches that can be promoted to splice at the transcriptional level in order to reconstitute a contiguous active HEase ORF when suitable levels of Mg^{+2} are present in the media [184,185].

Another regulation strategy involves the use of a destabilizing domain (DD) tag (12 kDa, 107 amino acid), which is based on a mutant of the FKBP12 protein (Fig. 2g) [214]. When the DD tag is attached to a protein of interest and expressed as a fusion protein, it leads to the rapid degradation of the protein in the cell by proteasomes. However, a protective effect is observed when the DD›s small (750 Da), membrane-permeant ligand (Shield-1) is added to the culture medium. This small ligand reversibly binds to the DD tag and protects the DD-tagged protein from degradation, leading to rapid accumulation of the tagged protein in the cell [215]. Previously, it was shown that linking a modified destabilizing FKBP12 (i.e. DD tag) domain to the amino-terminus of a ZFN protein destabilized the enzyme. A small molecule that blocks the destabilization effect of the amino-terminal domain was used to regulate the ZFN levels and this helped in maintaining higher rates of ZFN-mediated gene targeting while reducing genotoxicity [216]. Recently, Senturk and coworkers have shown that by fusing the FKBP12-derived DD to Cas9 (DD-Cas9), conditional regulation of Cas9 protein stability using DD ligand (Shield-1) could be achieved. Cas9 stability was reversed 2 h following Shield-1 ligand withdrawal from the media; the Cas9 levels were noted to be negligible within 12 h [217].

The various strategies of unifying the split-Cas9 into an active enzyme, or harnessing the splicing reaction of the internal introns in order to yield a functional MN, have been impressive. However, to the best of our knowledge, these “inducible” systems have not been put into any clinical settings. A list of current regulatable genome editing tools has been provided in Supplementary Table 1.

A recent study showed that there are natural inhibitors for CRISPR/Cas9 [218]. These inhibitors can bind to the Cas9 protein and they appear to be encoded by mobile elements and probably have evolved as defense mechanism by phages to counteract the bacterial CRISPR based immune systems. Three families of proteinaceous type of inhibitors have been identified in *Neisseria meningitidis* (Nme) that can potentially be used in human cells as “off” switches against NmeCas9 based genome editing reagents [218]. This could be a seminal study that will lead to further explorations on isolating natural Cas9 inhibitors, and the genes that encode them. These genes that encode Cas9 inhibitors under the control of inducible promoters could be employed as “off” switches in genome editing protocols.

ALTERNATIVE STRATEGIES FOR DEVELOPING REGULATABLE GENOME EDITING REAGENTS

Currently, there is a lot of focus on protein-based genome manipulation reagents. However, there are some noteworthy developments in trying to use oligonucleotides as potential alternatives to protein-based genome manipulation tools or as components of such systems (such as the previously discussed TFO nucleases or PNA based application). There would be several advantages of using oligonucleotides such as (a) ease of oligonucleotide synthesis and sequence verification, (b) predictable Watson-Crick base pairing allows for easier design against target sequence and addressing off-target issues, (c) the modular nature of RNA domains/structures permits engineering of multifunctional molecules [219], (d) design of oligonucleotides that target almost any molecule can be achieved by in vitro selection [220–222], (e) thermally denatured oligonucleotides are generally easier to renature than proteins, (f) some oligonucleotides are functional in the absence of protein factors (additional factors can increase the likelihood of side reactions), and (g) oligonucleotides are less likely to elicit an immune response. Some disadvantages or challenges include: (a) they can be difficult to identify and/or validate target candidates (i.e. not all oligonucleotides can be engineered to be inserted at any position within a sequence), (b) combining different oligonucleotide domains may decrease their efficiency and/or activity [223], (c) can form tertiary interactions, which is currently not fully understood (i.e. off-targeting potentially an issue), (d) in some cases, can be toxic to the cell (e.g. if a ligand is required at beyond physiological concentrations), and (e) prone to nuclease degradation. The best examples for oligonucleotide-based systems used to manipulate gene expression currently are RNAi and self-cleaving hammerhead [224,225]. These reagents allow for targeted control of gene expression by promoting the removal of specific mRNAs from the cytoplasm. Considerable work is still needed to develop oligonucleotides-based systems for genomic manipulation/editing.

Numerous oligonucleotide molecules are currently studied that could be coopted into regulatory elements at the mRNA level, but only select examples will be mentioned. These will be used to illustrate the potential for oligonucleotides as components of genome editing reagents. More specifically, how oligonucleotides could be incorporated as regulatory elements within protein-based genome editing reagents to refine their activity and accuracy of target site recognition.

The Utility of Hammerhead Ribozymes and Engineered Variants

The hammerhead ribozyme (HHR), first seen in tobacco ringspot virus satellite RNA [226], is an example of small nucleolytic RNA molecules capable of self-cleavage (i.e. ribozymes) [227]. Other autocatalytic (self-cleaving type) small RNA molecules are twister, twister sister, pistol, and hatchet ribozyme [228,229]. HHRs are composed of a conserved central sequence with three radiating helical domains [230]. Natural HHRs are not true ribozymes as they are only capable of carrying out a single self-cleavage reaction. Synthetic HHRs have been engineered to overcome this by separating the HHR into two components: ribozyme (the part of the HHR which remains unchanged) and substrate (the target sequence that will be cleaved).

Since their discovery in 1986 [226], HHRs have been noted in all domains of life [231]and have been extensively studied and modified. Several aspects of HHRs make them attractive as scaffolds for the development of regulatory switches for genome editing reagents: (a) short sequence (~ 50 nt for an active HHR [232]; enables faster troubleshooting and optimizing, and low cost of synthesis, (b) catalytic activity does not require any protein factors, which can lead to side reactions, (c) can be used as a genome editing tool, and (d) can be designed to cleave two different targets [233]. Some challenges of using HHRs are: (a) the substrate/target needs to be single-stranded in order for the ribozyme to bind, (b) minimal HHRs require Mg^{+2}concentrations to be above 10 mM, which is significantly higher than physiological concentrations (~ 0.1 mM) [234], (c) requirement of 5′-UX-3′ sequence at cleavage site, where X can be either A, C, or U [235] limits substrate design (although the limitation is not severe as such sequence is common within a genome).

HHRs are typically used as on/off switches at the mRNA level [236]. They can be integrated in the 5′- or 3′-untranslated region (UTR) in constructs that express genome editing reagents, and self-cleavage could regulate processing of the mRNA. Depending on the organism, insertion of ribozymes in the 5′- or 3′-UTR can evoke different effects. In prokaryotes, a common strategy is to engineer a HHR into the 5′-UTR such that one of the stems (stem I) of the HHR is modified to sequester (through base pairing) the ribosome binding sites (RBS in bacteria) or other features required for initiating translation (i.e. Kozak sequence in eukaryotes). Upon self-cleavage, the ribozyme is removed and the RBS is exposed permitting ribosome access and subsequent protein synthesis. In eukaryotes, using a similar approach to turn on translation would be difficult because self-cleavage of the ribozyme would lead to the removal of the 5′-cap, or 3′-poly(A) tail if the ribozyme if insertion occurred in the 3′-UTR. Both the 5′-cap, and 3′-poly(A) tail, play an essential role in mRNA stability and translation. Although HHRs have been successfully engineered into 5′-UTR [236], the use of this region, in general, can be challenging as the formation of hairpin structures may impede translation [237]. An alternative is the use of the 3′-UTR [238]. Here, self-cleavage is typically used to destabilize the mRNA; ultimately leading to its decay (i.e. ribozyme activity turns off protein expression).

Alternative designs with regards to HHRs are oligonucleotides developed by Erdmann›s group [239]. They developed a "mirror-image" hammerhead ribozyme and deoxyribozymes (DNAzymes), termed Spiegelzymes®, enantiomers of the biological D-nucleic acids. The advantage of using L-nucleic acids is that they are less prone to nuclease activity while still being able to interact with D-nucleic acids [240]. This strategy of generating enantiomers or synthetic analogs provides an alternative to standard HHR type molecules, which while more readily accessible, are less stable in an in vivo environment [241].

Another application of the HHR backbone involves the development of temperature-sensitive HHRs. Here, the incorporation of a "RNA thermometer" provides control over HHR activity. By replacing a stem-loop (stem III) of a HHR with a temperature-sensitive hairpin, Saragliadis et al. [242] developed a regulatory element where self-cleavage of a HHR is controlled by changes with temperature; cleavage occurs at lower temperatures but is inhibited at higher temperatures due to denaturing of the temperature-sensitive hairpin required in the formation of a catalytically active HHR. When this element is incorporated within the 5′-UTR region of an mRNA the self-cleavage reaction liberates features of the mRNA needed for initiating the translation of the transcript. The advantage of temperature-based systems is that the switch portion of the regulatory element does not require a ligand (a core requirement for standard riboswitches), which can pose as a challenge as many natural ligands are not long-lived, particularly in an in vivo environment. Although temperature control eliminates the need for a ligand, it can pose as a challenge, especially in mammals, where body temperature is tightly regulated. Thus, the use of such temperature-sensitive hairpins may be limited to prokaryotes and simpler eukaryotes (such as *Saccharomyces*), or in vitro experiments where temperature conditions can be more easily controlled.

Utility of Riboswitches and Allosteric Ribozymes

In many cases, temporal control of protein expression is desired (like components of genome editing tools). Thus, the sole use of a ribozyme is not sufficient as the ribozyme will self-splice once it is properly folded. Therefore, the ribozyme activity needs to be controlled, for example by the integration of a riboswitch element within the ribozyme molecule.

Riboswitches are RNA elements that modulate mRNA expression through binding of a ligand, which is typically a small organic molecule or ion, to its aptamer domain [243]. Ligand binding causes a conformational change in another part of the RNA, referred to as the expression platform, and alters mRNA expression [244]. The effect can either be positive or negative (i.e. promote or inhibit expression). There are currently numerous known riboswitches which can bind to a wide range of ligands including natural and synthetic analogs [243]. Advantages of riboswitches include: (a) no additional proteins are required (which can be toxic and/or deplete vital cell resources), (b) regulation of gene expression is achieved without the use of heterologous gene expression systems, which is characteristic of protein-based systems, and (c) ligand can be administered directly (protein-based systems typically require a transcription effector to be expressed from a plasmid vector).

Recently, several naturally occurring ribozymes have been described that may provide avenues for engineering regulatory switches that could be incorporated into genome editing reagents. Lee et al. [245] noted, in *Clostridium difficile* strain 630, a gene where the expression is regulated by both a riboswitch (aptamer binds c-di-GMP) and a ribozyme (group I intron), which occurs in tandem. In the cyclic di-GMP (c-di-GMP) example, activation of the riboswitch causes a conformational change resulting in splicing of the ribozyme (group I intron) and placement of the ribosome binding site at optimal distance from the start codon. When the ligand (c-di-GMP) is not bound to the aptamer domain, alternative splicing occurs generating a truncated mRNA with no ribosome binding site, which results deficient gene expression. An interesting feature of this riboswitch-ribozyme is that even with correct splicing of the ribozyme, the riboswitch continues to regulate gene expression. This is accomplished by the notion that the RBS remains sequestered in the basal stem of the riboswitch under low ligand concentrations, which inhibits binding of the ribosome. The gene is only expressed when ligand concentrations are high. This allows for tighter control and limiting leaky splicing of the ribozyme. Such tight regulation of gene expression would be desirable in engineering programmable genome editing systems in order to achieve temporal control to reduce off-target activities.

The glmS ribozyme is an example of an oligonucleotide that is both a ribozyme and riboswitch. It demonstrates how the modular nature of RNA can be applied to regulate gene expression. The expression of the *glmS* gene, encoding for glutamine-fructose 6-phosphate transaminase that catalyzes the formation of glucosamine 6-phosphate (GlcN6P), is regulated by binding of the ligand, GlcN6P. The ligand functions as a coenzyme; binding of the ligand provides an amine which participates in general acid-base catalysis [246]. When this element is incorporated into an expression system, self-cleavage of the mRNA leads to nonsense mediated decay. In this example, the control of gene expression is through negative feedback (i.e. binding of the enzyme product negatively influences the expression the enzyme). This strategy can be applied to modulate expression of a gene of interest (such as a component of the genome editing system) by designing a riboswitch–ribozyme element that responds to increasing protein concentration. Thus, protein concentrations can potentially be regulated to remain at relatively low levels. This is important, especially if the genome editing reagent is toxic in high concentrations, associated with off-target activities, or severely affects cell viability.

Another strategy is to incorporate an aptamer domain into the ribozyme, rather than having the ligand-binding and ribozyme components as separate entities (as above examples). Allosteric ribozymes [247], or aptazymes [248], are synthetic ribozymes whose catalytic activity is modulated by ligand binding. Several examples of aptazymes have been successfully synthesized and shown to be functional in both prokaryotic and eukaryotic systems [236,249,250]. Again, these types of engineered ribozymes could be part of expression vectors that encode components of genome editing systems.

CONCLUSION

Genome editing reagents are being developed and employed at a rapid rate. In order to increase specificity and avoid or reduce toxicity issues due to off-target activities strategies are now being developed to provide temporal control over the DNA-cutting activities of genome editing tools. The review presented a variety of novel approaches that have been employed so far but it also highlights the tremendous potential that is offered by nucleic acid-based regulatory switches that could be incorporated into the expression vectors of genome editing reagents. The development of programmable genome editing tools along with the ability of controlling the temporal and spatial expression of such editing reagents promises to be a very active and challenging research area.

The following is the supplementary data related to this article

Competing interests

The authors have declared that no competing interests exist.

ACKNOWLEDGMENTS

This work is supported by a Discovery grant from the Natural Sciences and Engineering Research Council of Canada (NSERC) to G.H. We also would like to acknowledge funding support from the University of Manitoba Faculty of Science Graduate Award program (T.K.G.) and the University of Manitoba Faculty of Graduate Studies GETS Program (T.K.G. and A.W.).

REFERENCES

1. Kodym A, Afza R. Physical and chemical mutagenesis. Methods Mol Biol 2003;236: 189–204.
2. O'Brien TP, Frankel WN. Moving forward with chemical mutagenesis in the mouse. J Physiol 2004;554:13–21.
3. Chalker JM, Davis BG. Chemical mutagenesis: selective post-expression interconversion of protein amino acid residues. Curr Opin Chem Biol 2010;14(6):781–9.
4. Ivics Z, Li MA, Mátés L, Boeke JD, Nagy A, et al. Transposon-mediated genome manipulation in vertebrates. Nat Methods 2009;6(6):415–22.
5. Muñoz-López M, García-Pérez JL. DNA transposons: nature and applications in genomics. Curr Genomics 2010;11(2):115–28.
6. Roland R, Lena R, Wang W, Cadinanos J, Vassiliou G, et al. PiggyBac transposon mutagenesis: a tool for cancer gene discovery in mice. Science 2010;330(6007): 1104–7.
7. Frantz S. Studies reveal potential pitfalls of RNAi. Nat Rev Drug Discov 2003;2: 763–4.
8. Castanotto D, Rossi JJ. The promises and pitfalls of RNA-interference-based therapeutics. Nature 2009;457:426–33.
9. von Bubnoff A. Adenovirus vectors: promise and possible pitfalls. IAVI Rep 2012; 16(1):4–7.
10. Aiuti A, Luca B, Samantha S, Francesca F, Maria PC, et al. Lentiviral hematopoietic stem cell gene therapy in patients with Wiskott–Aldrich syndrome. Science 2013;341:1233151.
11. Jasin M. Genetic manipulation of genomes with rare-cutting endonucleases. Trends Genet 1996;12(6):224–8.
12. Porteus M. Seeing the light: integrating genome engineering with double-strand break repair. Nat Methods 2011;8:628–30.
13. Kim H, Kim JS. A guide to genome engineering with programmable nucleases. Nat Rev Genet 2014;15:321–34.
14. Moore JK, Haber JE. Cell cycle and genetic requirements of two pathways of nonhomologous end-joining repair of double-strand breaks in Saccharomyces cerevisiae. Mol Cell Biol 1996;16(5):2164–73.

15. Lieber MR. The mechanism of double-strand DNA break repair by the nonhomologous DNA end joining pathway. Annu Rev Biochem 2010;79:181–211.
16. Choulika A, Perrin A, Dujon B, Nicolas JF. Induction of homologous recombination in mammalian chromosomes by using the I-SceI system of Saccharomyces cerevisiae. Mol Cell Biol 1995;15:1968–73.
17. Li X, Heyer W-D. Homologous recombination in DNA repair and DNA damage tolerance. Cell Res 2008;18(1):99–113.
18. Moehle EA, Rock JM, Lee YL, Jouvenot Y, DeKelver RC, et al. Targeted gene addition into a specified location in the human genome using designed zinc finger nucleases. Proc Natl Acad Sci U S A 2007;104:3055–60.
19. Lee HJ, Kim E, Kim JS. Targeted chromosomal deletions in human cells using zinc finger nucleases. Genome Res 2010;20(1):81–9.
20. Söllü C, Pars K, Cornu TI, Thibodeau-Beganny S, Maeder ML, et al. Autonomous zinc-finger nuclease pairs for targeted chromosomal deletion. Nucleic Acids Res 2010;38(22):8269–76.
21. Gaj T, Gersbach CA, Barbas III CF. ZFN, TALEN and CRISPR/Cas-based methods for genome engineering. Trends Biotechnol 2013;31(7):397–405.
22. Lin FL, Sperle K, Sternberg N. Repair of double-stranded DNA breaks by homologous DNA fragments during transfer of DNA into mouse L cells. Mol Cell Biol 1990;10:113–9.
23. Elliott B, Jasin M. Repair of double-strand breaks by homologous recombination in mismatch repair-defective mammalian cells. Mol Cell Biol 2001;21:2671–82.
24. Stark JM, Pierce AJ, Oh J, Pastink A, Jasin M. Genetic steps of mammalian homologous repair with distinct mutagenic consequences. Mol Cell Biochem 2004; 24(21):9305–16.
25. Wang H, Yang H, Shivalila CS, Dawlaty MM, Cheng AW, et al. One-step generation of mice carrying mutations in multiple genes by CRISPR/Cas-mediated genome engineering. Cell 2013;153:910–8.
26. Yasue A, Mitsui SN, Watanabe T, Sakuma T, Oyadomari S, et al. Highly efficient targeted mutagenesis in one-cell mouse embryos

mediated by the TALEN and CRISPR/Cas systems. Sci Rep 2014;4:5705.

27. Urnov FD, Miller JC, Lee YL, Beausejour CM, Rock JM, et al. Highly efficient endogenous human gene correction using designed zinc-finger nucleases. Nature 2005; 435:646–51.
28. Howe SJ, Mansour MR, Schwarzwaelder K, Bartholomae C, Hubank M, et al. Insertional mutagenesis combined with acquired somatic mutations causes leukemogenesis following gene therapy of SCID-X1 patients. J Clin Invest 2008;118(9): 3143–50.
29. Li H, Haurigot V, Doyon Y, Li T, Wong SY, et al. In vivo genome editing restores haemostasis in a mouse model of haemophilia. Nature 2011;475(7355):217–21.
30. Zou J, Mali P, Huang X, Dowey SN, Cheng L. Site-specific gene correction of a point mutation in human iPS cells derived from an adult patient with sickle cell disease. Blood 2011;118(17):4599–608.
31. Yusa K, Rashid ST, Strick-Marchand H, Varela I, Liu PQ, et al. Targeted gene correction of α1-antitrypsin deficiency in induced pluripotent stem cells. Nature 2011; 478(7369):391–4.
32. Gao H, Smith J, Yang M, Jones S, Djukanovic V, et al. Heritable targeted mutagenesis in maize using a designed endonuclease. Plant J 2010;61:176–87.
33. Watanabe K, Breier U, Hensel G, Kumlehn J, Schubert I, et al. Stable gene replacement in barley by targeted double-strand break induction. J Exp Bot 2016;67(5): 1433–45.
34. Deredec A, Burt A, Godfray HC. The population genetics of using homing endonuclease genes in vector and pest management. Genetics 2008;179:2013–26.
35. Windbichler DA, Papathanos PA, Catteruccia F, Ranson H, Burt A, et al. Homing endonuclease mediated gene targeting in Anopheles gambiae cells and embryos. Nucleic Acids Res 2007;35:5922–33.
36. Saunders S, Cooke B, McColl K, Shine R, Peacock T. Modern approaches for the biological control of vertebrate pests: an Australian perspective. Biol Control 2010;52(3):288–95.

37. Chan Y-S, Naujoks DA, Huen DS, Russell S. Insect population control by homing endonuclease-based gene drive: an evaluation in Drosophila melanogaster. Genetics 2011;188:33–44.
38. Chan Y-S, Takeuchi R, Jarjour J, Huen DS, Stoddard BL, Russell S. The design and in vivo evaluation of engineered I-OnuI-based enzymes for HEG gene drive. PLoS One 2013;8(9):e74254.
39. Gantz VM, Bier E. The mutagenic chain reaction: a method for converting heterozygous to homozygous mutations. Science 2015;348(6233):442–4.
40. Hammond A, Galizi R, Kyrou K, Simoni A, Siniscalchi C, et al. A CRISPR-Cas9 gene drive system targeting female reproduction in the malaria mosquito vector Anopheles gambiae. Nat Biotechnol 2016;34:78–83.
41. Scharenberg AM, Stoddard BL, Monnat RJ, Nolan A (2016) Retargeting: an unrecognized consideration in endonuclease-based gene drive biology. 10.1101/089946.
42. Stoddard BL. Homing endonuclease structure and function. Q Rev Biophys 2005; 38(1):49–95.
43. Silva G, Poirot L, Galetto R, Smith J, Montoya G, et al. Meganucleases and other tools for targeted genome engineering: perspectives and challenges for gene therapy. Curr Gene Ther 2011;11(1):11–27.
44. Hafez M, Hausner G. Homing endonucleases: DNA scissors on a mission. Genome 2012;55(8):553–69.
45. Stoddard BL. Homing endonucleases from mobile group I introns: discovery to genome engineering. Mob DNA 2014;5:7.
46. Kim YG, Cha J, Chandrasegaran S. Hybrid restriction enzymes: zinc finger fusions to FokI cleavage domain. Proc Natl Acad Sci U S A 1996;93:1156–60.
47. Pabo CO, Peisach E, Grant RA. Design and selection of novel Cys2His2 zinc finger proteins. Annu Rev Biochem 2001;70:313–40.
48. Cathomen T, Joung JK. Zinc-finger nucleases: the next generation emerges. Mol Ther 2008;16(7):1200–7.
49. Carroll D. Genome engineering with zinc-finger nucleases. GSA 2011;188(4): 773–82.
50. Boch J, Scholze H, Schornack S, Landgraf A, Hahn S, et al. Breaking

the code of DNA binding specificity of TAL-type III effectors. Science 2009;326(5959):1509–12.

51. Christian M, Cermak T, Doyle EL, Schmidt C, Zhang F, et al. Targeting DNA doublestrand breaks with TAL effector nucleases. Genetics 2010;186(2):757–61.

52. Boch J. TALEs of genome targeting. Nat Biotechnol 2011;29(2):135–6.

53. Li T, Huang S, Jiang WZ, Wright D, Spalding MH, et al. TAL nucleases (TALNs): hybrid proteins composed of TAL effectors and FokI DNA-cleavage domain. Nucleic Acids Res 2011;39:359–72.

54. Bhaya D, Davison M, Barrangou R. CRISPR-Cas systems in bacteria and archaea: versatile small RNAs for adaptive defense and regulation. Annu Rev Genet 2011; 45:273–97.

55. Jinek M, Chylinski K, Fonfara I, Hauer M, Doudna JA, Charpentier E. A programmable dual-RNA-guided DNA endonuclease in adaptive bacterial immunity. Science 2012;337(6096):816–21.

56. Jiang W, Bikard D, Cox D, Zhang F, Marraffini LA. RNA-guided editing of bacterial genomes using CRISPR-Cas systems. Nat Biotechnol 2013;31(3):233–9.

57. Zimmerly S, Guo H, Perlman PS, Lambowitz AM. Group II intron mobility occurs by target DNA-primed reverse transcription. Cell 1995;82:545–54.

58. Lambowitz AM, Zimmerly S. Group II introns: mobile ribozymes that invade DNA. Cold Spring Harb Perspect Biol 2011;3(8):a003616.

59. Mohr G, Hong W, Zhang J, Cui GZ, Yang Y. A targetron system for gene targeting in thermophiles and its application in Clostridium thermocellum. PLoS One 2013; 8(7):e69032.

60. Nisa-Martínez R, Molina-Sánchez MD, Toro N. Host factors influencing the retrohoming pathway of group II intron RmInt1, which has an intron-encoded protein naturally devoid of endonuclease activity. PLoS One 2016;11(9):e0162275.

61. Mohr G, Smith D, Belfort M, Lambowitz AM. Rules for DNA target-site recognition by a lactococcal group II intron enable retargeting of the intron to specific DNA sequences. Genes Dev 2000;14(5):559–73.

62. Karberg M, Guo H, Zhong J, Coon R, Perutka J, Lambowitz AM. Group II introns as controllable gene targeting vectors for genetic manipulation of bacteria. Nat Biotechnol 2001;19:1162–7.
63. Cui X, Davis G. Mobile group II intron targeting: applications in prokaryotes and perspectives in eukaryotes. Front Biosci 2007;12:4972–85.
64. Wyman C, Kanaar R. DNA double-strand break repair: all's well that ends well. Annu Rev Genet 2006;40:363–83.
65. Ciccia A, Elledge SJ. The DNA damage response: making it safe to play with knives. Mol Cell 2010;40:179–204.
66. Chapman JR, Taylor MR, Boulton SJ. Playing the end game: DNA double-strand break repair pathway choice. Mol Cell 2012;47:497–510.
67. Cristea S, Freyvert Y, Santiago Y, Holmes MC, Urnov FD, et al. In vivo cleavage of transgene donors promote nuclease-mediated targeted integration. Biotechnol Bioeng 2013;110:871–80.
68. Maresca M, Lin VG, Guo N, Yang Y. Obligate ligation-gated recombination (ObLiGaRe): custom-designed nuclease-mediated targeted integration through nonhomologous end joining. Genome Res 2013;23:539–46.
69. Giles SH, Yeo I, Sadaf F, Shiva A, David JH, et al. A frameshift mutation in MC4R associated with dominantly inherited human obesity. Nat Genet 1998;20:111–2.
70. Ogura Y, Bonen DK, Inohara N, Nicolae DL, Chenn FF, et al. A frameshift mutation in NOD2 associated with susceptibility to Crohn's disease. Nature 2001;411:603–6.
71. You KT, Li LS, Kim N-G, Kang HJ, Koh KH, et al. Selective translational repression of truncated proteins from frameshift mutation-derived mRNAs in tumors. PLoS Biol 2007;5(5):e109.
72. Ousterout DG, Perez-Pinera P, Thakore PI, Kabadi AM, Brown MT, et al. Reading frame correction by targeted genome editing restores dystrophin expression in cells from Duchenne muscular dystrophy patients. Mol Ther 2013;21: 1718–26.
73. Li HL, Fujimoto N, Sasakawa N, Shirai S, Ohkame T, et al. Precise correction of the dystrophin gene in Duchenne muscular dystrophy

patient induced pluripotent stem cells by TALEN and CRISPR-Cas9. Stem Cell Rep 2015;4(1):143–54.

74. Lin S, Staahl BT, Alla RK, Doudna JA. Enhanced homology-directed human genome engineering by controlled timing of CRISPR/Cas9 delivery. Elife 2014;3:e04766.

75. Cai Y, Laustsen A, Zhou Y, Sun C, Anderson MV, et al. Targeted, homology-driven gene insertion in stem cells by ZFN-loaded 'all-in-one' lentiviral vectors. Elife 2016;5:e12213.

76. Sander JD, Joung JK. CRISPR-Cas systems for editing, regulating and targeting genomes. Nat Biotechnol 2014;32(4):347–55.

77. Hsu PD, Lander ES, Zhang F. Development and applications of CRISPR-Cas9 for genome engineering. Cell 2014;157(6):1262–78.

78. Enyeart PJ, Mohr G, Ellington AD, Lambowitz AM. Biotechnological applications of mobile group II introns and their reverse transcriptases: gene targeting, RNA-seq, and non-coding RNA analysis. Mob DNA 2014;5:21.

79. Cox DBT, Platt RJ, Zhang F. Therapeutic genome editing: prospects and challenges. Nat Med 2015;21(2):121–31.

80. Porteus MH. Towards a new era in medicine: therapeutic genome editing. Genome Biol 2015;16:286.

81. Maeder ML, Gersbach CA. Genome-editing technologies for gene and cell therapy. Mol Ther 2016;24(3):430–46.

82. Kim JS. Genome editing comes of age. Nat Protoc 2016;11:1573–8.

83. Kim E, Kim S, Hyoung-Kim D, Choi B-S, Choi IK-Y, et al. Precision genome engineering with programmable DNA-nicking enzymes. Genome Res 2012;22:1327–33.

84. Calos MP. Genome editing techniques and their therapeutic applications. Clin Pharmacol Ther 2016. http://dx.doi.org/10.1002/cpt.542.

85. Singh A, Chakraborty D, Maiti S. CRISPR/Cas9: a historical and chemical biology perspective of targeted genome engineering. Chem Soc Rev 2016. http://dx.doi. org/10.1039/c6cs00197a.

86. Osborn MJ, Belanto JJ, Tolar J, Voytas DF. Gene editing and its application for hematological diseases. Int J Hematol 2016;104:18–28.

87. Wang L, Li F, Dang L, Liang C, Wang C, et al. In vivo delivery systems for therapeutic genome editing. Int J Mol Sci 2016;17(5).
88. Egholm M, Buchardt O, Christensen L, Behrens C, Freier SM, et al. PNA hybridizes to complementary oligonucleotides obeying the Watson–Crick hydrogen bonding rules. Nature 1993;365(6446):566–8.
89. Simon P, Cannata F, Concordet JP, Giovannangeli C. Targeting DNA with triplexforming oligonucleotides to modify gene sequence. Biochimie 2008;90(8):1109–16.
90. Varshney GK, Burgess SM. DNA-guided genome editing using structure-guided endonucleases. Genome Biol 2016;17(1):187.
91. Xu S, Cao S, Zou B, Yue Y, Gu C, et al. An alternative novel tool for DNA editing without target sequence limitation: the structure-guided nuclease. Genome Biol 2016; 17(1):186.
92. Algasaier SI, Exell JC, Bennet IA, Thompson MJ, Gotham VJ, et al. DNA and protein requirements for substrate conformational changes necessary for human flap endonuclease-1-catalyzed reaction. J Biol Chem 2016;291(15):8258–68.
93. Grizot S, Epinat J-C, Thomas S, Duclert A, Rolland S, et al. Generation of redesigned homing endonucleases comprising DNA-binding domains derived from two different scaffolds. Nucleic Acids Res 2010;38(6):2006–18.
94. Moure CM, Gimble FS, Quiocho FA. Crystal structure of the intein homing endonuclease PI-SceI bound to its recognition sequence. Nat Struct Biol 2002;9(10): 764–70.
95. Ashworth J, Taylor GK, Havranek JJ, Quadri SA, Stoddard BL, et al. Computational reprogramming of homing endonuclease specificity at multiple adjacent base pairs. Nucleic Acids Res 2010;38(16):5601–8.
96. Belfort M, Bonocora RP. Homing endonucleases: from genetic anomalies to pro-grammable genomic clippers. Methods Mol Biol 2014;1123:1–26.
97. Jacoby K, Lambert AR, Scharenberg AM. Characterization of homing endonuclease binding and cleavage specificities using yeast

surface display SELEX (YSD-SELEX). Nucleic Acids Res 2016. http://dx.doi.org/10.1093/nar/gkw864.

98. Kanchiswamy CN, Maffei M, Malnoy M, Velasco R, Kim J-S. Fine-tuning nextgeneration genome editing tools. Trends Biotechnol 2016;34:7.
99. Takeuchi R, Lambert AR, Nga-Sze Mak A, Jacoby K, Dickson RJ, et al. Tapping natural reservoirs of homing endonucleases for targeted gene modification. Proc Natl Acad Sci U S A 2011;108(32):13077–82.
100. Taylor GK, Petrucci LH, Lambert AR, Baxter SK, Jarjour J, et al. LAHEDES: the LAGLIDADG homing endonuclease database and engineering server. Nucleic Acids Res 2012;40(Web Server issue):W110–6. http://dx.doi.org/10.1093/nar/gks365.
101. Roberts RJ, Vincze T, Posfai J, Macelis D. REBASE: restriction enzymes and methyltransferases. Nucleic Acids Res 2003;31(1):418–20.
102. Lambert AR, Hallinan JP, Shen BW, Chik JK, Bolduc JM, et al. Indirect DNA sequence recognition and its impact on nuclease cleavage activity. Structure 2016;24(6): 862–73.
103. Boissel S, Jarjour J, Astrakhan A, Adey A, Gouble A, et al. megaTALs: a rare-cleaving nuclease architecture for therapeutic genome engineering. Nucleic Acids Res 2014; 42(4):2591–601.
104. Li MV, Shukla D, Rhodes BH, Lall A, Shu J, et al. HomeRun vector assembly system: a flexible and standardized cloning system for assembly of multi-modular DNA constructs. PLoS One 2014;9(6):e100948.
105. Liu JK, Chen WH, Ren SX, Zhao GP, Wang J. iBrick: a new standard for iterative assembly of biological parts with homing endonucleases. PLoS One 2014; 9:e110852.
106. McVey M, Lee SE. MMEJ repair of double-strand breaks (director's cut): deleted sequences and alternative endings. Trends Genet 2008;24:529–38.
107. Certo MT, Gwiazda KS, Kuhar R, Sather B, Curinga G, et al. Coupling endonucleases with DNA end-processing enzymes to drive gene disruption. Nat Methods 2012;9: 973–5.

108. Wolfs JM, DaSilva M, Meister SE, Wang X, Schild-Poulter C, et al. MegaTevs: singlechain dual nucleases for efficient gene disruption. Nucleic Acids Res 2014;42(13): 8816–29.

109. Wolfs JM, Hamilton TA, Lant JL, Laforet M, Zhang J, et al. Biasing genome-editing events toward precise length deletions with an RNA-guided TevCas9 dual nuclease. Proc Natl Acad Sci U S A 2016. http://dx.doi.org/10.1073/pnas.1616343114.

110. Urnov FD, Rebar EJ, Holmes MC, Zhang HS, Gregory PD. Genome editing with engineered zinc finger nucleases. Nat Rev Genet 2010;11(9):636–46.

111. Miller JC, Holmes MC, Wang J, Guschin DY, Lee Y-L, et al. An improved zinc-finger nuclease architecture for highly specific genome editing. Nat Biotechnol 2007;25: 778–85.

112. Szczepek M, Brondani V, Büchel J, Serrano L, Segal DJ, et al. Structure-based redesign of the dimerization interface reduces the toxicity of zinc-finger nucleases. Nat Biotechnol 2007;25:786–93.

113. Didigu CA, Wilen CB, Wang J, Duong J, Secreto AJ, et al. Simultaneous zinc-finger nuclease editing of the HIV coreceptors ccr5 and cxcr4 protects CD4+ T cells from HIV-1 infection. Blood 2014;123(1):61–9.

114. Tebas P, Stein D, Tang WW, Frank I, Wang SQ, et al. Gene editing of CCR5 in autologous CD4 T cells of persons infected with HIV. N Engl J Med 2014;370(10): 901–10.

115. Pattanayak V, Ramirez CL, Joung JK, Liu DR. Revealing off-target cleavage specificities of zinc-finger nucleases by in vitro selection. Nat Methods 2011;8: 765–70.

116. Mino T, Mori T, Aoyama Y, Sera T. Gene- and protein-delivered zinc finger staphylococcal nuclease hybrid for inhibition of DNA replication of human papillomavirus. PLoS One 2013;8(2):e56633.

117. Mino T, Mori T, Aoyama Y, Sera T. Inhibition of DNA replication of human papillomavirus by using zinc finger-single-chain FokI dimer hybrid. Mol Biotechnol 2014; 56(8):731–7.

118. Miller JC, Tan S, Qiao G, Barlow KA, Wang J, et al. A TALE nuclease architecture for efficient genome editing. Nat Biotechnol 2011;29:143–8.

119. Kim Y, Kweon J, Kim JS. TALENs and ZFNs are associated with different mutation signatures. Nat Methods 2013;10:185.
120. Doyle EL, Booher NJ, Standage DS, Voytas DF, Brendel VP, et al. TAL EffectorNucleotide Targeter (TALE-NT) 2.0: tools for TAL effector design and target prediction. Nucleic Acids Res 2012;40(Web Server issue):W117–22. http://dx.doi.org/10. 1093/nar/gks608.
121. Hockemeyer D, Wang H, Kiani S, Lai CS, Gao Q, et al. Genetic engineering of human pluripotent cells using TALE nucleases. Nat Biotechnol 2011;29:731–4.
122. Suzuki K, Yu C, Qu J, Li M, Yao X, et al. Targeted gene correction minimally impacts whole-genome mutational load in human-disease-specific induced pluripotent stem cell clones. Cell Stem Cell 2014;15(1):31–6.
123. Wang X, Wang Y, Wu X, Wang J, Wang Y, et al. Unbiased detection of off-target cleavage by CRISPR-Cas9 and TALENs using integrase-defective lentiviral vectors. Nat Biotechnol 2015;33:175–8.
124. Kleinstiver BP, Wang L, Wolfs JM, Kolaczyk T, McDowell B, et al. The I-TevI nuclease and linker domains contribute to the specificity of monomeric TALENs. G3 (Bethesda) 2014;4(6):1155–65.
125. Barrangou R. CRISPR-Cas systems and RNA-guided interference. Wiley Interdiscip Rev RNA 2013;4(3):267–78.
126. Ran FA, Hsu PD, Wright J, Agarwala V, Scott DA, Zhang F. Genome engineering using the CRISPR-Cas9 system. Nat Protoc 2013;8:2281–308.
127. McCaffrey J, Sibert J, Zhang B, Zhang Y, Hu W, et al. CRISPR-CAS9 D10A nickase target-specific fluorescent labeling of double strand DNA for whole genome mapping and structural variation analysis. Nucleic Acids Res 2016;44(2):e11. http://dx. doi.org/10.1093/nar/gkv878.
128. Dickinson DJ, Goldstein B. CRISPR-based methods for Caenorhabditis elegans genome engineering. Genetics 2016;202(3):885–901.
129. Karvelis T, Gasiunas G, Miksys A, Barrangou R, Horvath P, et al. crRNA and tracrRNA

130. guide Cas9-mediated DNA interference in Streptococcus thermophilus. RNA Biol 2013;10(5):841–51.
131. Hou Z, Zhang Y, Propson NE, Howden SE, Chu LF, et al. Efficient genome engineering in human pluripotent stem cells using Cas9 from Neisseria meningitidis. Proc Natl Acad Sci U S A 2013;110(39):15644–9.
132. Kleinstiver BP, Prew MS, Tsai SQ, Topkar VV, Nguyen NT, et al. Engineered CRISPRCas9 nucleases with altered PAM specificities. Nature 2015;523(7561):481–5.
133. Qi LS, Larson MH, Gilbert LA, Doudna JA, Weissman JS. Repurposing CRISPR as an
134. RNA-guided platform for sequence-specific control of gene expression. Cell 2013; 152(5):1173–83.
135. Gasiunas G, Barrangou R, Horvath P, Siksnys V. Cas9–crRNA ribonucleoprotein complex mediates specific DNA cleavage for adaptive immunity in bacteria. Proc Natl Acad Sci U S A 2012;109(39):E2579–86.
136. Maeder ML, Linder SJ, Cascio VM, Fu Y, Ho QH, et al. CRISPR RNA-guided activation of endogenous human genes. Nat Methods 2013;10(10):977–9.
137. Mali P, Aach J, Stranges PB, Esvelt KM, Moosburner M, Kosuri S, et al. CAS9 transcriptional activators for target specificity screening and paired nickases for cooperative genome engineering. Nat Biotechnol 2013;31:833–8.
138. Xu T, Li Y, Van Nostrand JD, He Z, Zhou J. Cas9-based tools for targeted genome editing and transcriptional control. Appl Environ Microbiol 2014;80(5):1544–52.
139. Young SA, Aitken RJ, Ikawa M. Advantages of using the CRISPR/Cas9 system of genome editing to investigate male reproductive mechanisms using mouse models. Asian J Androl 2015;17(4):623–7.
140. Lackner DH, Carré A, Guzzardo PM, Banning C, Mangena R, et al. A generic strategy for CRISPR-Cas9-mediated gene tagging. Nat Commun 2015;6:10237.

141. Zalatan JG, Lee ME, Almeida R, Gilbert LA, Whitehead EH, et al. Engineering complex synthetic transcriptional programs with CRISPR RNA scaffolds. Cell 2015; 160:339–50.

142. Didovyk A, Borek B, Tsimring L, Hasty J. Transcriptional regulation with CRISPR-Cas9: principles, advances, and applications. Curr Opin Biotechnol 2016;40:177–84.

143. Mandegar MA, Huebsch N, Frolov EB, Shin E, Truong A, et al. CRISPR interference efficiently induces specific and reversible gene silencing in human iPSCs. Cell Stem Cell 2016;18:541–53.

144. Chen B, Gilbert LA, Cimini BA, Schnitzbauer J, Zhang W, et al. Dynamic imaging of genomic loci in living human cells by an optimized CRISPR/Cas system. Cell 2013;155(7):1479–91.

145. Tsai SQ, Wyvekens N, Khayter C, Foden JA, Thapar V, et al. Dimeric CRISPR RNAguided FokI nucleases for highly specific genome editing. Nat Biotechnol 2014; 32:569–76.

146. Cong L, Ran FA, Cox D, Lin S, Barretto R, et al. Multiplex genome engineering using CRISPR/Cas systems. Science 2013;339:819–23.

147. Platt RJ, Chen S, Zhou Y, Yim MJ, Swiech L, et al. CRISPR-Cas9 knockin mice for genome editing and cancer modeling. Cell 2014;159(2):440–55.

148. Ousterout DG, Kabadi AM, Thakore PI, Majoros WH, Reddy TE, et al. Multiplex CRISPR/Cas9-based genome editing for correction of dystrophin mutations that cause Duchenne muscular dystrophy. Nat Commun 2015;6:6244.

149. Kaur K, Tandon H, Gupta AK, Kumar M. CrisprGE: a central hub of CRISPR/Casbased genome editing. Database (Oxford) 2015:bav055. http://dx.doi.org/10. 1093/database/bav055.

150. Calos MP. The CRISPR way to think about Duchenne's. N Engl J Med 2016;374: 1684–6.

151. Peng R, Lin G, Li J. Potential pitfalls of CRISPR/Cas9-mediated genome editing. FEBS J 2016;283(7):1218–31.

152. Ledford H. CRISPR, the disruptor. Nature 2015;522:20–4.

153. Ran FA, Hsu PD, Lin CY, Gootenberg JS, Konermann S, et al. Double nicking by RNAguided CRISPR Cas9 for enhanced genome editing specificity. Cell 2013;154: 1380–9.

154. Shen B, Zhang W, Zhang J, Zhou J, Wang J, et al. Efficient genome modification by CRISPR-Cas9 nickase with minimal off-target effects. Nat Methods 2014;11: 399–402.

155. Fu Y, Sander JD, Reyon D, Cascio VM, Joung JK. Improving CRISPR-Cas nuclease specificity using truncated guide RNAs. Nat Biotechnol 2014;32(3):279–84.

156. Esvelt KM, Mali P, Braff JL, Moosburner M, Yaung SJ, et al. Orthogonal Cas9 proteins for RNA-guided gene regulation and editing. Nat Methods 2013;10:1116–21.

157. Zetsche B, Gootenberg JS, Abudayyeh OO, Slaymaker IM, Makarova KS. Cpf1 is a single RNA-guided endonuclease of a class 2 CRISPR-Cas system. Cell 2015;163(3): 759–71.

158. Li SY, Zhao GP, Wang J. C-Brick: a new standard for assembly of biological parts using Cpf1. ACS Synth Biol 2016. http://dx.doi.org/10.1021/acssynbio.6b00114.

159. Fonfara I, Richter H, Bratovič M, Rhun AL, Charpentier E. The CRISPR-associated DNA-cleaving enzyme Cpf1 also processes precursor CRISPR RNA. Nature 2016; 532:517–21.

160. Kleinstiver BP, Tsai SQ, Prew MS, Nguyen NT, Welch MM, et al. Genome-wide specificities of CRISPR-Cas Cpf1 nucleases in human cells. Nat Biotechnol 2016;34(8): 869–74.

161. Kim D, Kim J, Hur JK, Been KW, Yoon SH, Kim JS. Genome-wide analysis reveals specificities of Cpf1 endonucleases in human cells. Nat Biotechnol 2016;34(8): 863–8.

162. Ran FA, Cong L, Yan WX, Scott DA, Gootenberg JS, et al. In vivo genome editing using Staphylococcus aureus Cas9. Nature 2015;520:186–91.

163. Slaymaker IM, Gao L, Zetsche B, Scott DA, Yan WX, et al. Rationally engineered Cas9 nucleases with improved specificity. Science 2015;351(6268):84–8.

164. Komor AC, Yongjoo BK, Packer MS, Zuris JA, Liu DR. Programmable editing of a target base in genomic DNA without double-stranded DNA cleavage. Nature 2016; 533:420–4.

165. Harris RS, Petersen-Mahrt SK, Neuberger MS. RNA editing enzyme APOBEC1 and some of its homologs can act as DNA mutators. Mol

Cell 2002;10:1247–53.

166. Nelles DA, Fang MY, O'Connell MR, Xu JL, Markmiller SJ, et al. Programmable RNA tracking in live cells with CRISPR/Cas9. Cell 2016;165(2):488–96.
167. Knauert MP, Glazer PM. Triplex forming oligonucleotides: sequence-specific tools for gene targeting. Hum Mol Genet 2001;10(20):2243–51.
168. Abibi A, Protozanova E, Demidov VV, Frank-Kamenetskii MD. Specific versus nonspecific binding of cationic PNAs to duplex DNA. Biophys J 2004;86(5):3070–8.
169. Anstaett P, Gasser G. Peptide nucleic acid - an opportunity for bio nanotechnology. Chimia 2014;68(4):264–8.
170. Pingoud A, Jeltsch A. Structure and function of type II restriction endonucleases. Nucleic Acids Res 2001;29:3705–27.
171. Pingoud A, Fuxreiter M, Pingoud V, Wende W. Type II restriction endonucleases: structure and mechanism. Cell Mol Life Sci 2005;62:685–707.
172. Silanskas A, Zaremba M, Sasnauskas G, Siksnys V. Catalytic activity control of restriction endonuclease-triplex forming oligonucleotide conjugates. Bioconjug Chem 2012;23(2):203–11.
173. Devi G, Zhou Y, Zhong Z, Toh DF, Chen G. RNA triplexes: from structural principles to biological and biotech applications. Wiley Interdiscip Rev RNA 2012;6(1):111–28.
174. Gao F, Shen XZ, Jiang F, Wu Y, Han C. DNA-guided genome editing using the Natronobacterium gregoryi Argonaute. Nat Biotechnol 2016;34(7):768–73.
175. Cyranoski D. Replications, ridicule and a recluse: the controversy over NgAgo geneediting intensifies. Nature 2016;536(7615):136–7.
176. Thyagarajan B, Olivares EC, Hollis RP, Ginsburg DS, Calos MP. Site-specific genomic integration in mammalian cells mediated by phage phiC31 integrase. Mol Cell Biol 2001;21:3926–34.
177. Bertoni C, Jarrahian S, Wheeler TM, Li Y, Olivares EC, et al. Enhancement of plasmid-mediated gene therapy for Duchenne muscular dystrophy by directed plasmid integration. Proc Natl Acad Sci U S A 2006;103:419–24.

178. Sauer B, Henderson N. Cre-stimulated recombination at loxP containing DNA sequences placed into the mammalian genome. Nucleic Acids Res 1989;17:147–61.
179. Groth AC, Olivares EC, Thyagarajan B, Calos MP. A phage integrase directs efficient site-specific integration in human cells. Proc Natl Acad Sci U S A 2000;97:5995–6000.
180. Grindley NDF, Whiteson KL, Rice PA. Mechanisms of site specific recombination. Annu Rev Biochem 2006;75:567–605.
181. Chalberg TW, Portlock JL, Olivares EC, Thyagarajan B, Kirby PJ, et al. Integration specificity of phage phiC31 integrase in the human genome. J Mol Biol 2006;357:28–48.
182. Chavez CL, Calos MP. Therapeutic applications of the phiC31 integrase system. Curr Gene Ther 2011;11:375–81.
183. Zhao C, Farruggio AP, Bjornson CRR, Chavez CL, Geisinger JM, et al. Recombinasemediated reprogramming and dystrophin gene addition in mdx mouse induced pluripotent stem cells. PLoS One 2014;9(4):e96279.
184. Posey KL, Gimble FS. Insertion of a reversible redox switch into a rare-cutting DNA endonuclease. Biochemistry 2002;41(7):2184–90.
185. Guha TK, Hausner G. A homing endonuclease with a switch: characterization of a twintron encoded homing endonuclease. Fungal Genet Biol 2014;65:57–68.
186. Guha TK, Hausner G. Using group II introns for attenuating the in vitro and in vivo expression of a homing endonuclease. PLoS One 2016;11(2):e0150097.
187. Guha TK, Hausner G. Insertion of ribozyme based switches into homing endonucle-ase genes. In: Reeves A, editor. Methods in molecular biology—in vitro mutagene-sis: methods and protocols. Springer Verlag; 2016. p. 135–52.
188. Dai L, Toor N, Olson R, Keeping A, Zimmerly S. Database for mobile group II introns. Nucleic Acids Res 2003;31(1):424–6.
189. Candales MA, Duong A, Hood KS, Li T, Neufeld RA, et al. Database for bacterial group II introns. Nucleic Acids Res 2012;40(Database issue):D187–90.

190. Abebe M, Candales MA, Duong A, Hood KS, Li T. A pipeline of programs for collecting and analyzing group II intron retroelement sequences from GenBank. Mob DNA 2013;4:28.
191. Bonen L. Cis- and trans-splicing of group II introns in plant mitochondria. Mitochondrion 2008;8:26–34.
192. Merendino L, Perron K, Rahire M, Howald I, Rochaix JD, et al. A novel multifunctional factor involved in trans-splicing of chloroplast introns in Chlamydomonas. Nucleic Acids Res 2006;34:262–74.
193. Quiroga C, Kronstad L, Ritlop C, Filion A, Cousineau B. Contribution of base pairing interactions between group II intron fragments during trans-splicing in vivo. RNA 2011;17:2212–21.
194. Mastroianni M, Watanabe K, White TB, Zhuang F, Vernon J, et al. Group II intronbased gene targeting reactions in eukaryotes. PLoS One 2008;3(9):e3121.
195. Truong DM, Sidote DJ, Russell R, Lambowitz AM. Enhanced group II intron retrohoming in magnesium-deficient E. coli via selection of mutations in the ribo-
196. zyme core. Proc Natl Acad Sci U S A 2013;110:3800–9.
197. Truong DM, Hewitt FC, Hanson JH, Cui X, Lambowitz AM. Retrohoming of a mobile group II intron in human cells suggests how eukaryotes limit group II intron proliferation. PLoS Genet 2015;11:e1005422.
198. Cho SW, Lee J, Carroll D, Kim J-S, Lee J. Heritable gene knockout in Caenorhabditis elegans by direct injection of Cas9-sgRNA ribonucleoproteins. Genetics 2013; 195(3):1177–80.
199. Kim S, Kim D, Cho SW, Kim J, Kim J-S. Highly efficient RNA-guided genome editing in human cells via delivery of purified Cas9 ribonucleoproteins. Genome Res 2014; 24(6):1012–9.
200. Yoshioka S, Fujii W, Ogawa T, Sugiura K, Naito K. Development of a monopromoter-driven CRISPR/Cas9 system in mammalian cells. Sci Rep 2015;5:18341.
201. Cao J, Wu L, Zhang SM, Lu M, Cheung WKC, et al. An easy and efficient inducible CRISPR/Cas9 platform with improved specificity for multiple gene targeting. Nucleic Acids Res 2016. http://dx.doi.org/10.1093/nar/gkw660.

202. Wright AV, Sternberg SH, Taylor DW, Staahl BT, Bardales JA, et al. Rational design of a split-Cas9 enzyme complex. Proc Natl Acad Sci U S A 2015;112(10):2984–9.
203. Dow LE, Fisher J, O'Rourke KP, Muley A, Kastenhuber ER, et al. Inducible in vivo genome editing with CRISPR-Cas9. Nat Biotechnol 2015;33:390–4.
204. Zetsche B, Volz SE, Zhang F. A split-Cas9 architecture for inducible genome editing and transcription modulation. Nat Biotechnol 2015;33:139–42.
205. Laplante M, Sabatini DM. mTOR signaling in growth control and disease. Cell 2012; 149:274–93.
206. Schierling B, Noël AJ, Wende W, Hien le T, Volkov E, et al. Controlling the enzymatic activity of a restriction enzyme by light. Proc Natl Acad Sci U S A 2010;107(4): 1361–6.
207. Zaremba M, Siksnys V. Molecular scissors under light control. Proc Natl Acad Sci U S A 2010;107(4):1259–60.
208. Brieke C, Rohrbach F, Gottschalk A, Mayer G, Heckel A. Light-controlled tools. Angew Chem Int Ed Engl 2012;51:8446–76.
209. Nihongaki Y, Kawano F, Nakajima T, Sato M. Photoactivatable CRISPR-Cas9 for optogenetic genome editing. Nat Biotechnol 2015;33(7):755–60.
210. Kawano I, Suzuki H, Furuya A, Sato M. Engineered pairs of distinct photoswitches for optogenetic control of cellular proteins. Nat Commun 2015;6:6256.
211. Zuris JA, Thompson DB, Shu Y, Guilinger JP, Bessen JL. Cationic lipid-mediated delivery of proteins enables efficient protein-based genome editing in vitro and in vivo. Nat Biotechnol 2015;33(1):73–80.
212. Goodsell DS. The molecular perspective: ultraviolet light and pyrimidine dimers. Oncologist 2001;6(3):298–9.
213. Sinha RP, Häder DP. UV-induced DNA damage and repair: a review. Photochem Photobiol Sci 2002;1(4):225–36.
214. Davis KM, Pattanayak V, Thompson DB, Zuris JA, Liu DR. Small molecule triggered Cas9 protein with improved genome-editing specificity. Nat Chem Biol 2015;11: 316–8.

215. Liu KI, Ramli MNB, Woo CWA, Wang Y, Zhao T, et al. A chemical-inducible CRISPR-Cas9 system for rapid control of genome editing. Nat Chem Biol 2016;12: 980–7.
216. Guilinger JP, Thompson DB, Liu DR. Fusion of catalytically inactive Cas9 to FokI nuclease improves the specificity of genome modification. Nat Biotechnol 2014; 32(6):577–82.
217. Banaszynski LA, Chen L, Maynard-Smith LA, Ooi AGL, Wandless TJ. A rapid, reversible, and tunable method to regulate protein function in living cells using synthetic small molecules. Cell 2006;126(5):995–1004.
218. Armstrong CM, Goldberg DE. An FKBP destabilization domain modulates protein levels in Plasmodium falciparum. Nat Methods 2007;4:1007–9.
219. Pruett-Miller SM, Reading DW, Porter SN, Porteus MH. Attenuation of zinc finger nuclease toxicity by small-molecule regulation of protein levels. PLoS Genet 2009;5(2):e1000376.
220. Senturk S, Shirole NH, Nowak DD, Corbo V, Vaughan A, et al. (2015) A rapid and tunable method to temporally control Cas9 expression enables the identification of essential genes and the interrogation of functional gene interactions in vitro and in vivo. 10.1101/023366.
221. Pawluk A, Amrani N, Zhang Y, Garcia B, Hidalgo-Reyes Y, et al. Naturally occurring offswitches for CRISPR-Cas9. Cell 2016. http://dx.doi.org/10.1016/j.cell.2016.11.017.
222. Grabow W, Jaeger L. RNA modularity for synthetic biology. F1000Prime Rep 2013; 5:46.
223. Ellington AD, Szostak JW. In vitro selection of RNA molecules that bind specific ligands. Nature 1990;346(6287):818–22.
224. Tuerk C, Gold L. Systematic evolution of ligands by exponential enrichment: RNA ligands to bacteriophage T4 DNA polymerase. Science 1990;249(4968):505–10.
225. Li Y. A quarter century of in vitro selection. J Mol Evol 2015;81(5–6):137–9.
226. Link KH, Breaker RR. Engineering ligand-responsive gene-control elements: lessons learned from natural riboswitches. Gene Ther 2009;16(10):1189–201.

227. Li Z, Rana TM. Therapeutic targeting of microRNAs: current status and future challenges. Nat Rev Drug Discov 2014;13(8):622–38.

228. Scott WG, Horan LH, Martick M. The hammerhead ribozyme: structure, catalysis, and gene regulation. Prog Mol Biol Transl Sci 2013;120:1–23.

229. Prody GA, Bakos JT, Buzayan JM, Schneider IR, Bruening G. Autolytic processing of

230. dimeric plant virus satellite RNA. Science 1986;231(4745):1577–80.

231. Müller S, Appel B, Balke D, Hieronymus R, Nübel C. Thirty-five years of research into ribozymes and nucleic acid catalysis: where do we stand today? F1000Res 2016;5.

232. Roth A, Weinberg Z, Chen AG, Kim PB, Ames TD, et al. A widespread self-cleaving ribozyme class is revealed by bioinformatics. Nat Chem Biol 2014;10(1):56–60.

233. Weinberg Z, Kim PB, Chen TH, Li S, Harris KA, et al. New classes of self-cleaving ribozymes revealed by comparative genomics analysis. Nat Chem Biol 2015; 11(8):606–10.

234. Ruffner DE, Stormo GD, Uhlenbeck OC. Sequence requirements of the hammerhead RNA self-cleavage reaction. Biochemistry 1990;29(47):10695–702.

235. Hammann C, Luptak A, Perreault J, de la Peña M. The ubiquitous hammerhead ribozyme. RNA 2012;18(5):871–85.

236. O'Rourke SM, Estell W, Scott WG. Minimal hammerhead ribozymes with uncompromised catalytic activity. J Mol Biol 2015;427(14):2340–7.

237. Iyo M, Kawasaki H, Taira K. Maxizyme technology. Methods Mol Biol 2004;252: 257–65.

238. Khvorova A, Lescoute A, Westhof E, Jayasena SD. Sequence elements outside the hammerhead ribozyme catalytic core enable intracellular activity. Nat Struct Biol 2003;10(9):708–12.

239. Simorre JP, Legault P, Baidya N, Uhlenbeck OC, Maloney L, et al. Structural variation induced by different nucleotides at the cleavage site of the hammerhead ribozyme. Biochemistry 1998;37(12):4034–44.

240. Beilstein K, Wittmann A, Grez M, Suess B. Conditional control of mammalian gene expression by tetracycline-dependent hammerhead ribozymes. ACS Synth Biol 2015;4(5):526–34.
241. Pelletier J, Sonenberg N. Insertion mutagenesis to increase secondary structure within the 5′ noncoding region of a eukaryotic mRNA reduces translational efficiency. Cell 1985;40(3):515–26.
242. Win MN, Smolke CD. A modular and extensible RNA-based gene-regulatory platform for engineering cellular function. Proc Natl Acad Sci U S A 2007;104(36): 14283–8.
243. Wyszko E, Szymański M, Zeichhardt H, Müller F, Barciszewski J, Erdmann VA. Spiegelzymes: sequence specific hydrolysis of L-RNA with mirror image hammerhead ribozymes and DNAzymes. PLoS One 2013;8(1):e54741.
244. Wyszko E, Mueller F, Gabryelska M, Bondzio A, Popenda M, et al. Spiegelzymes® mirror-image hammerhead ribozymes and mirror-image DNAzymes, an alternative to siRNAs and microRNAs to cleave mRNAs in vivo? PLoS One 2014;9(1):e86673.
245. Griffin LC, Tidmarsh GF, Bock LC, Toole JJ, Leung LL. In vivo anticoagulant properties of a novel nucleotide-based thrombin inhibitor and demonstration of regional anticoagulation in extracorporeal circuits. Blood 1993;81(12):3271–6.
246. Saragliadis A, Krajewski SS, Rehm C, Narberhaus F, Hartig JS. Thermozymes: synthetic RNA thermometers based on ribozyme activity. RNA Biol 2013;10(6): 1010–6.
247. Machtel P, Bąkowska-Żywicka K, Żywicki M. Emerging applications of riboswitches — from antibacterial targets to molecular tools. J Appl Genet 2016. http://dx.doi.org/10.1007/s13353-016-0341-x.
248. Batey RT. Riboswitches: still a lot of undiscovered country. RNA 2015;21(4):560–3.
249. Lee ER, Baker JL, Weinberg Z, Sudarsan N, Breaker RR. An allosteric self-splicing ribozyme triggered by a bacterial second messenger. Science 2010;329(5993):845–8.
250. Klein DJ, Ferré-D'Amaré AR. Structural basis of glmS ribozyme activation by glucosamine-6-phosphate. Science 2006;313(5794):1752–6.

251. Porta H, Lizardi PM. An allosteric hammerhead ribozyme. Biotechnology (N Y) 1995;13(2):161–4.
252. Thompson KM, Syrett HA, Knudsen SM, Ellington AD. Group I aptazymes as genetic regulatory switches. BMC Biotechnol 2002;2:21.
253. Ogawa A, Maeda M. An artificial aptazyme-based riboswitch and its cascading system in E. coli. Chembiochem 2008;9(2):206–9.
254. Wieland M, Hartig JS. Improved aptazyme design and in vivo screening enable riboswitching in bacteria. Angew Chem Int Ed Engl 2008;47(14):2604–7.

7 A HYBRID COMPUTER-AIDED-DIAGNOSIS SYSTEM FOR PREDICTION OF BREAST CANCER RECURRENCE (HPBCR) USING OPTIMIZED ENSEMBLE LEARNING

Mohammad R. Mohebian [a], Hamid R. Marateb [a,b], Marjan Mansourian [c],Miguel Angel Mañanas [b], Fariborz Mokarian [d,e]

[a] Biomedical Engineering Department, Engineering Faculty, University of Isfahan, Hezar Jerib St., 81746-73441, Isfahan, Iran

[b] Department of Automatic Control, Biomedical Engineering Research Center, Universitat Politècnica de Catalunya, BarcelonaTech (UPC), C. Pau Gargallo, 5, 08028 Barcelona, Spain

[c] Department of Biostatistics and Epidemiology, School of Public Health, Isfahan University of Medical Sciences, Hezar Jerib St., 81745 Isfahan, Iran

[d] Cancer Prevention Research Center, Isfahan University of Medical Sciences, Isfahan, Iran

[e] Department of Internal Medicine, School of Medicine, Isfahan University of Medical Sciences, Isfahan, Iran

ABSTRACT

Cancer is a collection of diseases that involves growing abnormal cells with the potential to invade or spread to the body. Breast cancer is the second leading cause of cancer death among women. A method for 5-year breast cancer recurrence prediction is presented in this manuscript. Clinicopathologic characteristics of 579 breast cancer patients (recurrence prevalence of 19.3%) were analyzed and discriminative features were selected using statistical feature selection methods. They were further refined by Particle Swarm Optimization (PSO) as the inputs of the classification system with ensemble learning (Bagged Decision Tree: BDT). The proper combination of selected categorical features and also the weight (importance) of the selected interval-measurement-scale features were identified by the PSO algorithm. The performance of HPBCR (hybrid predictor of breast cancer recurrence) was assessed using the holdout and 4-fold cross-validation. Three other classifiers namely as supported vector machines, DT, and multilayer perceptron neural network were used for comparison. The selected features were diagnosis age, tumor size, lymph node involvement ratio, number of involved axillary lymph nodes, progesterone receptor expression, having hormone therapy and type of surgery. The minimum sensitivity, specificity, precision and accuracy of HPBCR were 77%, 93%, 95% and 85%, respectively in the entire cross-validation folds and the hold-out test fold. HPBCR outperformed the other tested classifiers. It showed excellent agreement with the gold standard (i.e. the oncologist opinion after blood tumor marker and imaging tests, and tissue biopsy). This algorithm is thus a promising online tool for the prediction of breast cancer recurrence.

INTRODUCTION

Computer-aided diagnosis (CAD) is using computers and software to interpret medical information. The purpose of CAD is to improve the diagnosis accuracy. In fact, CAD is used as a second opinion by the physicians to make the final diagnosis decision [1,2].

Nowadays, CAD is used in many different fields in medicine including, but not limited to, early detection of breast cancer [3], lung cancer diagnosis [4], arrhythmia detection [5] and dental and maxillofacial lesions diagnosis [6]. Several studies have been reported in the literature focusing on the application of CAD for cancer diagnosis and prognosis [7,8].

Cancer is a collection of diseases that involves growing abnormal cells with the potential to spread to other parts of the body [9]. There are over 200 types of cancer. The most common types of cancer in women are breast, colorectal, lung, and cervical [10].

Cancer is the leading cause of death worldwide, accounting for 8.2 million deaths in 2012 [11]. The most common causes of cancer death are: lung (1.59 million deaths), liver (745,000), stomach (723,000), colorectal (694,000), breast (521,000), and esophageal (400,000) [11]. More than 60% of world's total new annual cancer cases occur in Africa, and Central and South America [12]. The cancer incidence in Asia is also high [13]. It is expected that annual cancer cases will rise from 14 million in 2012 to 22 million within the next 2 decades [11,12]. Worthwhile, among all types of cancer, breast cancer is the second leading cause of cancer death among women [14].

In Iran, cancer is the third cause of death after coronary heart disease, and accidents [15]. Breast cancer is the leading type of cancer in Iranian females, accounting for 24.6% of all cancers. In Iran, the average women age with breast cancer is 49.6 years [16]. Among all of the provinces in Iran, Isfahan Province is the biggest and most important area located at a desert border in the center of Iran [17]. The rate of cancer is increasing rapidly in Isfahan Province. Cancer control and comprehensive prevention plan is thus necessary [18].

Breast cancer, a complex heterogeneous disease, occurs with a set of different clinical symptoms. It is usually diagnosed using blood tests, MRI, mammography, and CT scan and biopsy. The pathology results from biopsy samples indicate whether the suspicious area is cancerous. Cancer patients, then, undergo a systematic treatment procedure dependent on the cancer stage and additional lab tests such as hormone receptor status [19].

Cancer staging (assigning an ordinal numbers I–IV) is in fact the process to determine the extent to which a cancer has been spread and is based on the following four characteristics: 1) the tumor size, 2) whether the cancer is invasive or non-invasive, 3) the spread of the tumor into the lymph nodes, and 4) the spread of the tumor into other parts of the body (i.e. metastasis). Stage I is an isolated cancer (tumor size ≤ 20 mm in greatest dimension) while stage IV is a metastasis cancer. Most cancer deaths are due to cancer that has spread from its primary site to other organs [12,20].

Breast cancer is usually treated with surgery, which may be followed by chemotherapy, radiation, and hormone therapies [21]. When the cancer patient is initially treated, the disease can recur at any time. However, most recurrences happen in the first 5 years after treatment [22]. The recurrence could be local (near the mastectomy scar), regional (spread to nearby lymph nodes) or metastatic (spread to other parts of the body not near the breast). Some of the most common sites of recurrence outside the breast are the lymph nodes, bones, liver, lungs, and brain [23].

An important issue is whether we can optimize the treatments to increase the therapeutic efficacy. In fact, 5-year recurrence-free survival is an important treatment quality measure. In principle, it is possible to predict 5-year cancer recurrence using clinic pathologic characteristics of cancer patients [24]. Such a prediction could be used by doctors to make proper treatment plan to considerably prolong patient life [25].

The prediction systems were used for cancer diagnosis in the literature [7]. However, there are few studies focusing on cancer prognosis (including recurrence or survival analysis). Since the focus of the current study is prediction of cancer recurrence, the literature review on cancer recurrence prediction models is provided. Meanwhile, the table of available methods was provided

Zeng suggested a mixture classification model containing a two-layer structure called mixture of rough set and support vector machine (SVM) for breast cancer prognosis with the average accuracy of 91% [26].

The Nottingham prognostic index (NPI) is a prognostic regression model proposed by Galea et al. [27] based on tumor size, histological grade, and lymph node status. The NPI calculation equation is as follows: tumor size (cm) $\times$ 0.2 + histological grade + lymph node point (negative nodes = 1; 1–3 positive nodes = 2; $\geq$ 4 positive nodes = 3). The patients were then classified into the low-risk (NPI point $<$ 3.4) and high-risk groups (NPI point $\geq$ 3.4) [27,28].

Kim et al. [28] used normalized mutual information index for feature selection and supported vector machines (SVM), Cox-proportional hazard regression model, and artificial neural network classifiers for classification in a sample size of 679 patients (the recurrence prevalence of 28.6%). The following features were used in their prognosis system: local invasion of tumor, number of tumors, number of metastatic lymph nodes, the histological grade, tumor size, estrogen receptor, and lymphovascular invasion and reached the sensitivity, specificity and area under the curve of 89%, 73% and 0.85, respectively for the best classifier (SVM). Although the statistical power of their system is acceptable (Power = 89% > 80%), the Type I error is beyond the acceptable range ($\alpha = 0.17 > 0.05$).

Ahmad et al. [29] used the SVM, decision tree, and multilayer perceptron artificial neural network classifiers with feature selection in a sample size of 547 patients (the recurrence prevalence of 21.4%). The predictors were age at diagnosis, menarche and menopause, tumor size, number of involved and dissected axially lymph nodes, grade and HER2. The best classifier (SVM) had sensitivity, specificity and accuracy of 96%, 91% and 94%, respectively. Despite the high performance of the algorithm proposed by the authors, cases with primary metastasis, a metastasis diagnosed at the time of registry, were not excluded from the analysis. In fact, having metastasis was one of the attributes primarily used by their proposed system. This necessary exclusion has been implemented in similar studies [28,30]. The reason for such an exclusion is that the survival rate of patients with stage IV (metastasis) breast cancer within 5 years is about 10% to 15%. Thus, most of patients with primary metastasis experience cancer recurrence.

The discriminative prognosis predictors vary regionally in different studies [27–29]. Also, such prognosis tools have been recently used to optimize the treatment protocol. An example is including the cancer recurrence risk to breast cancer treatment guidelines by the American Society of Clinical Oncology (ASCO) and the National Comprehensive Cancer Network (NCCN). Thus, there is a need to design prognosis tools in regions that are different based on cancer epidemiology. Meanwhile, it is necessary to validate the cancer prognosis system using extensive diagnosis validation criteria. Therefore, the aim of this study was to *reliably* and *accurately* predict breast cancer recurrence using pathological and demographics features of the patients. In the proposed CAD system, HPBCR (hybrid predictor of breast cancer recurrence), a hybrid technique including statistical features selection, meta-heuristic population-based optimization and ensemble learning were used to predict breast cancer recurrence in the first 5 years after the diagnosis.

The rest of the paper is organized as follows: in the next section, information about the experimental protocol and the pattern recognition methods used in this study is presented. Section 3 provides the results of HPBCR and its comparison with the state-of-the-art. The discussion is provided in Section 4 and finally, the conclusions are summarized in Section 5.

MATERIALS AND METHODS

The Cohort Dataset

In this study we used a 16-year registry cohort database (1998–2014) on 1085 women who diagnosed with breast cancer in Isfahan Sayed-o-Shohada cancer research center. History of clinical conditions and therapy of patients was continued until death or lost to follow up. Characteristics of variables and tumor have been recorded by interview and reported as pathology results. Time to recurrence was based on the physicians' opinion. The following information was extracted from each patient: age at diagnosis of breast cancer, lymph node involvement ratio (NR) defined as ratio of involved to dissected lymph nodes [31], age of menarche, number of pregnancy (No. Preg), primary tumor size (TS), cellular marker for proliferation (Ki67), number of involved (I. Node) and dissected (T. Node) axillary lymph nodes, number of chemotherapies (No. Chemo) and categorical covariates of family history of cancer (FH: negative/positive), having more than one tumor in the breast (multifocal: negative/positive), estrogen receptor expression (ER: negative (also known as absent)/positive), progesterone receptor expression (PR: negative (also known as absent)/positive), tumor protein 53 (p53: negative/positive), type of surgery (MRM: modified radical mastectomy, BCS: breast-conserving surgery, Mast: mastectomy), epidermal growth factor receptor-2 (Her2: negative/positive (also known as overexpressed)), Cathepsin-D protein status (Cathepsin: negative/positive), hormone therapy (HRT: negative/positive), radiotherapy (XRT: negative/positive) and the histological grade of tumor (1 to 4). Moreover, five main molecular subtypes of breast cancer were extracted as shown to affect recurrence prognosis in the literature. They were calculated using ER, PR, HER2 and Ki-67 (the cutoff value of 14% to define 'low' and 'high') status. These subtypes are: luminal A (ER and/or PR-positive/HER2 negative/low Ki-67), luminal B (ER- and/or PR-positive/HER2-negative/high Ki-67), HER2-positive luminal B (ER- and/or PR-positive/HER2 positive/any Ki-67), non-luminal HER2-positive (ER and PR negative/HER2 positive) and triple negative (ER and PR negative/HER2-negative) [32].

Eighty nine subjects were excluded from the analysis because of primary metastasis, a metastasis diagnosed at the time of registry [30]. A number of 579 breast cancer patients with complete information were involved in our analysis. All subjects gave informed consent to the experimental procedure. The experimental protocol was approved by the Isfahan University of Medical Sciences Panel on Medical Human Subjects and conformed to the Declaration of Helsinki.

The structure of HPBCR is depicted in Fig. 1. First, the statistical features' selection was used to identify significant discriminative features. Among such features, categorical features (i.e. nominal or ordinal measurement scales [33]) used in the Bagged Decision Tree (BGD) were selected by Particle Swarm Optimization (PSO). The features with the interval measurement scale used by BGD were assigned weights being also tuned by PSO. The fitness function of the PSO was the F-score of the classifier on the training set (Eq. (1)). In the following, the details HPBCR are discussed.

$$fitness = FScore_{train} \qquad (1)$$

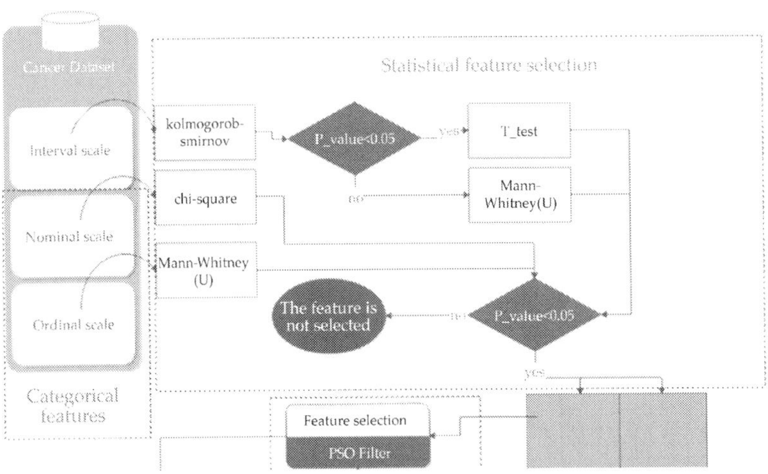

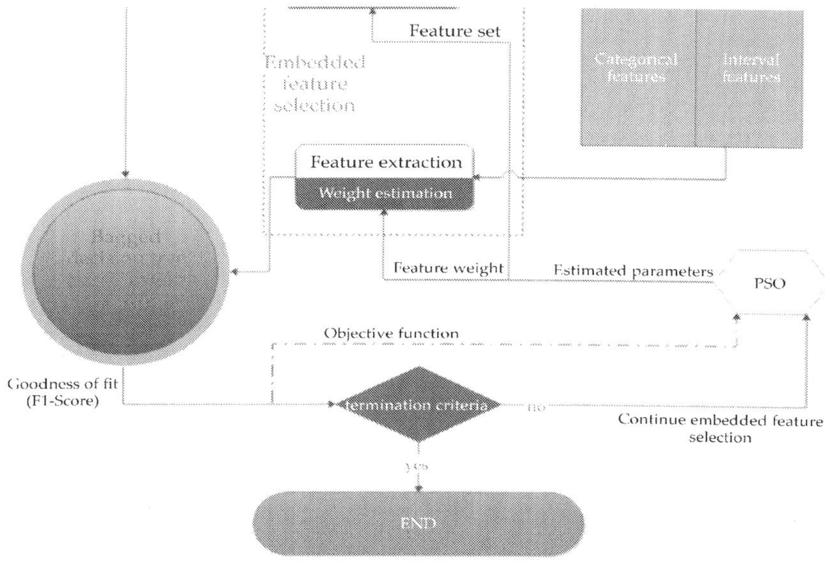

Figure 1: The structure of the proposed prognosis system (HPBCR). Other classifiers such as SVM and MLP could be used instead of BDT. The pseudo-code of HPBCR is provided in the. The input features in HPBCR were: diagnosis age, nodal ratio, menarche age, the number of pregnancy, tumor size, Ki67, the number of involved and dissected nodes, as interval features, type of surgery, molecular subtypes, family history of cancer, multifocal tumor, estrogen and progesterone receptor status, p53 mutation, Her2 expression, Cathepsin-D protein status, using hormone therapy, and radiotherapy as nominal features and the tumor grade as an ordinal feature. Briefly, the input features are first selected using statistical feature selection. The selected features were used by Bagged Decision Tree to build the classifier. The optimal feature set and the weight of features are estimated using Particle Swarm Optimization during learning. The algorithm stops if no significant improvement is seen in the objective function or the maximum number of iterations (set to 100 in our study) is reached.

Statistical Feature Selection

Feature selection (FS), the process of selecting a subset of relevant features, is divided into wrapper, embedded and filter methods. Filter method selects subsets of variables as a pre-processing step, independently from the chosen classifier while the other two methods, are used during machine learning procedure [34]. In wrapper methods, the selection of a features set is formulated as a search problem, in which different feature combinations are evaluated. Embedded methods, on the other hand, learn which set best contribute to the model accuracy while the model is being created. The former method searches the entire feature subsets while in the later one, search is guided by learning process which is prone to less over-fitting [35].

In fact, Filter FS also known as Statistical FS (SFS) applies a statistical measure to assign a score to each feature [36]. Such features are then ranked and selected based on the resulting score. Univariate SFS was first used in our analysis. Accordingly, the normality of the features with interval measurement scale was assessed using the Kolmogorov–Smirnov (KS) test [37]. Then, independent-sample t-test was used to identify statistically discriminative normally-distributed features. Meanwhile, Wilcoxon Mann–Whitney test was used to distinguish discriminative features with ordinal or non-normal interval measurement scales [38] (Fig. 1).

Chi-square test was used to check whether a nominal-measurement-scale feature was discriminative. The discriminative features were selected and further refined as the inputs of the classification system (BDT), next discussed.

Bagged Decision Tree

Decision tree builds classification models in the form of a tree structure. It divides a dataset into small subsets to incrementally develop an associated decision tree with decision and leaf nodes [39]. It uses entropy to calculate the homogeneity of samples to build the tree [40]. We used the statistical classifier C4.5 decision tree [41] with pruning in HPBCR. C4.5 sorts the data at every node of the tree to determine the best splitting attribute. When the tree is constructed, pruning is carried out from the leaves to the root. The redundant sub-trees are removed and thus the prediction error decreases. C4.5 handles both continuous and discrete attributes and is thus suitable for our application. Moreover, the online implementation of the resulting tree is simple.

Decision tree has been applied to many fields; for example in prediction of breast cancer relapse after surgery [42], classification mass spectrometry of blood serum samples from people with and without lung cancer [43] and gene expression data for cancer classification [44]. In fact, decision trees are reliable and effective decision making techniques used in different areas of medical decision making [45]. Meanwhile, in evidence-based medicine, it was used for clinical decision analysis [46].

Bagging is a method for improving results of machine learning classification algorithms. This method was formulated by Breiman and its name was deduced from the phrase "bootstrap aggregating". The details of bagging could be found in [47,48]. Bagging is usually applied to decision tree methods, though it can be used with other classifiers. In summary bagged tree, resamples the training dataset several times (bootstrapping), and builds a decision tree model from each; then aggregate these models together for a final classifier. This method was used for breast cancer classification [49], gene expression data analysis [50] and prostate cancer classification [51]. In our study, the number of trees was set to 10 for bagging.

The proper combination of categorical selected features and also the weight (importance) of the selected interval-measurement-scale features were identified using a meta-heuristics population-based stochastic optimization method entitled as Particle Swarm Optimization (PSO) (Fig. 1), discussed in the next chapter.

Particle Swarm Optimization (PSO)

PSO is an evolutionary computational method inspired by flocking birds [52], applied in many different areas, including manufacturing [53], civil engineering [54], optimum design [55], computational neuroscience [56] and breast cancer detection [57]. PSO is initialized with a population of random solutions known as particles, growing over generations to find optimal solutions. In a population, each particle has a velocity (i.e. rate of change in solutions) to enable them to fly through the problem space. Therefore, each particle is represented by a position and a velocity [58]. The current velocity of a particle depends on its previous velocity and on the distances of the particle from the personal (cognitive term) and neighborhood (social term) best positions (Eq. (2)).

$$v_i^{n+1} = wv_i^n + c_1 r_1^n \left(p_i^n - x_i^n\right) + c_2 r_2^n \left(p_g^n - x_i^n\right) \tag{2}$$

where w is the inertia weight; c_1, c_2 are the cognitive and social acceleration coefficients; i is the particle number, r_1 and r_2 are independent random numbers whose elements are uniformly distributed in [0, 1]; n denotes the iteration number, p_i is the personal best of the ith particle and p_g is the neighborhood best.

Each particle records its best position (personal best) so far and the best position achieved in the group (neighborhood best) among all personal bests. Then, the new position of a particle is determined based on its previous position information and its current velocity (Eq. (3)).

$$x_i^{n+1} = x_i^n + v_i^{n+1} \quad (3)$$

The objective is to find the particle that best maximizes the target fitness function (Eq. (1)). Moreover, the neighborhood of each particle is the entire swarm in our method (star topology).

In HPBCR, the acceleration coefficients c_1 and c_2 were set to 2. A large value of inertia weight supports global search ("exploration"), while a small value favors local search ("exploitation"). A suitable strategy is to first motivate exploration, and then exploitation [56]. Thus, the inertia coefficient was set to 1.00 at the first iteration and was linearly decreased with the damping coefficient of 0.99 at each iteration, and the number of particles was set to 20 and 100 iterations were performed. Moreover, in order to increase the convergence possibility, velocity clamping (V_{max} = 4) was used.

PSO algorithm was used to filter categorical features and to estimate the weights of interval features (Fig. 1). In categorical features, the position of the particles was rounded to 0 and 1 (Binary PSO) but for interval features, such values were real numbers between zero and one. The pseudo-code of the embedded feature selection based on PSO was provided in the (section B).

State-of-the-art

In our study, other classification methods namely as MLP, SVM and DT were used for comparison. Such methods were proposed in similar studies in the literature [28,29]. The selected features using SFS were used as the input to these classifiers.

MLP is a feed-forward artificial neural network model that maps sets of input data onto a set of appropriate outputs. It consists of multiple fully connected layers of nodes in which learning happens by changing connection weights through back propagation. It is a modification of the standard linear perceptron and can be used for classification of data that are not linearly separable [59]. In this study the MLP with 1 hidden layer with 10 neurons and sigmoid activation function was used. Such an activation function was reported by Isa et al. to be suitable for the prediction of breast cancer [60]. Meanwhile, the sensitivity analysis was performed on the number of neurons. Cross-validation on the training set (70% of data) was performed. When increasing the number of neurons up to 10, the prediction accuracy increased and did not improve considerably afterwards. Thus, ten neurons were used as to avoid over-fitting problem.

SVM constructs a hyper plane or set of hyper planes in a high or infinite-dimensional space, which can be used for classification [61]. The SVM algorithm has been widely applied in the biological and many other fields [62–64]. In our study, the radial basis function (RBF) kernels were used. The soft-margin parameter and the radius of the RBF kernel should be set properly, because inappropriate parameter settings result in poor classification. The method proposed by Wu and Wang [65] was used to set the soft-margin and RBF kernel parameters.

Validation

The Performance Indices for Each Classifier

The performance of the classifiers was assessed using the holdout method, an approach to out-of-sample evaluation, where the dataset was split into two mutually exclusive sets (70% training and 30% test sets). The classifiers were then trained on the first set and tested on the other set [66]. Additionally, 4-fold cross-validation was used to assess the performance of the best classifier to overcome a possible biased error estimate [67]. A comprehensive list of performance indices [68–72] was selected and reported for each classifier to identify different statistical errors which is crucial in medical diagnosis especially in cancer studies. The performance measures of the classifiers are listed in Table 1, along with their definitions. Discriminant power (DP) was characterized as "poor" when $DP < 1$, "limited" when $DP < 2$, "fair" when $DP < 3$ and "good" in other cases [72,73]. Among those measures AUC ROC is a global measure of diagnosis accuracy. ROC (receiver operating characteristic) curve is a plot of true positive rate (i.e. sensitivity) versus false positive rate (i.e. 1-specificity) of the diagnosis system as its discrimination threshold is varied. AUC (area under roc curve) is an effective and combined measure of the validity of the diagnosis test, which is also known as balanced accuracy defined the average of the sensitivity and specificity of the diagnosis system [72]. Its ranges for excellent, very good and good diagnosis accuracy are (0.9–1.0), (0.8–0.9) and (0.7–0.8) respectively [74].

Table 1: The classification performance measures used in our study.

Se = TPTP+FN	Sp = TNTN+FP	Acc = TP+TNTP+TN+FP+FN
Pr = TPTP+FP	Alpha = 1 − Sp	Beta = 1 − Se
F-score = 2×Pr×SePr+Se	AUC = 12×Sp+Se	LR+=Se1−Sp
LR−=1−SeSp	DOR=LR+LR−	
MCC = TP×TN−FP×FNTP+FP×TP+FN×TN+FP×TN+FN		
DP = 3π×logDOR		

True positive (TP): subjects with cancer recurrence, correctly identified; false positive (FP): subjects without recurrence, incorrectly identified; true negative (TN): subjects without recurrence, correctly identified; false negative (FN): subjects with recurrence, incorrectly identified; Se: sensitivity = Power; Sp: specificity; Acc: Accuracy; Pr: precision; AUC: area under the receiver operating characteristic (ROC) curve; LR: likelihood ratio; DOR: diagnosis odds ratio; MCC: Matthews correlation coefficient; DP: discriminant power.

Along with such indices, the agreement rate between the output of the classifier and the gold standard was assessed using Cohen's kappa coefficient [75]. Kappas over 0.75 was characterized as excellent, 0.40 to 0.75 as fair to good, and below 0.40 as poor [76]. A diagnosis system was considered as reliable if (1) its sensitivity and specificity were at least 80% (the minimum statistical power of 80%) and 95% (the maximum Type I error of 0.05), respectively [77,78], and (2) its false discovery rate (FDR = 1-precision) is as low as 5% [79]. Moreover, as a general rule, diagnosis tests with positive likelihood ratio (LR^+) greater than 10 or negative likelihood ratio (LR^-) less than 0.1 are preferred [80]. The application of sensitivity and specificity is highly dependent on the prevalence of the disease. A conservative method could be fulfilling both LR conditions (i.e. DOR greater than 100).

Comparison Between Different Classifiers

McNemar's test, also known as the Gillick test, was used to compare different classifiers to identify whether one classifier statistically significantly outperforms the other or not [67,81]. It was originally used to analyze retrospective case–control studies, where each case is matched to a particular control or it can be used to analyze experimental studies, where two treatments are given to match subjects [82].

To compare two classifiers A and B, *z* statistics is estimated as:

$$z = \frac{|x - y| - 1}{\sqrt{x + y}} \quad (4)$$

where x is the number of samples misclassified by A but not B and y is number of samples misclassified by B but not A. The null hypothesis (that the classifiers have the same error) can be rejected (with probability of incorrect rejection of 0.05) if $|z| > 1.96$ [67,83]. In such a case, a classifier with higher DP value was reported as superior to the other one.

All data processing was performed off-line using Matlab version 8.6 (The MathWorks Inc., Natick, MA, USA). All statistical analysis and calculations were performed using the SPSS statistical package, version 18.0 (SPSS Inc., Chicago, IL, USA).

Online Web-based HPBCR Implementation

Ten extracted rules were obtained from BDT whose feature weights were optimized using PSO. The prognosis system was written in C# language and used in ASP.net5 which is an open-source framework from Microsoft Corporation. The prediction model is freely available in the website (http://www.Prognosis.ir).

RESULTS

The average age of the participants was 46.8 ± 10.0 years. Among the number of 579 patients participated in our study, 19.3% had cancer recurrence during five years after diagnosis. The demographic and pathological characteristics of the participants, grouped by their classification with/without recurrence, are depicted in Tables 2, 3 and 4 for the features with interval, nominal and ordinal measurement scales, respectively. In fact for the features with binary, nominal with more than two categories, ordinal and interval measurement scales, their existence, category, rank and value is important in the diagnosis system. Among the interval measurement scale features, only age at diagnosis and TS were normally distributed in the recurrence group. Thus, due to the rejection of class dependence normality, Mann–Whitney test was used for SFS in interval features. Selected discriminative features using SFS were: age at diagnosis, NR, TS, I. Node, T. Node, Ki67, PR, HRT, surgery and cancer subtypes.

Table 2: Comparison of clinical and biochemical features (with interval measurement scale) of included subjects with/without cancer recurrence in $\mu \pm \sigma$ [min, max].

Variable	With recurrence	Without recurrence
Age*(year)	45.4 ± 10.1 (ND)[27.0, 68.0]	47.2 ± 9.9[24, 75.0]
NR*	0.47 ± 0.36[0.0, 1.0]	0.24 ± 0.32[0.0, 1.0]
Menarche	13.3 ± 1.3[11.0, 17.0]	13.4 ± 1.4[10.0, 17.0]
No. Preg	3.5 ± 1.9[0.0, 10.0]	3.7 ± 2.1[0.0, 10.0]
TS*(cm)	4.0 ± 1.7 (ND)[1.0, 8.5]	3.7 ± 1.8[0.5, 11.0]
Ki67*	21.1 ± 12.6[6.0, 74.0]	18.0 ± 11.8[1.0, 86.0]
I. Node*	6.6 ± 7.3[0.0, 39.0]	3.0 ± 4.9[0.0, 36.0]
T. Node*	12.8 ± 6.1[1.0, 39.0]	11.4 ± 5.4[1.0, 36.0]
No. Chemo	7.7 ± 0.9[4.0, 10.0]	7.4 ± 1.3[1.0, 9.0]

Age (age at diagnosis); NR (lymph node involvement ratio); menarche (menarche age); No. Preg (number of pregnancy); TS (tumor size); Ki67 (Ki67 proliferation marker); I. Node: number of involved auxiliary lymph nodes; T. Node: number of dissected auxiliary lymph nodes; No. Chemo: number of chemotherapy; ND: normally distributed. The sample size was 579 and the recurrence prevalence was 19.3%. Statistically significant features ($P < 0.05$).

Table 3: Comparison of clinical and biochemical features (with nominal measurement scale) of included subjects with/without cancer recurrence in percentage.

Variable	With recurrence	Without recurrence
FH	P 18N: 82	P: 22 N: 78
Multifocal	33 67	27 73
ER	50 50	57 43
PR*	50 50	61 39
P53	42 58	34 66
Surgery*	MRM: 51 BCS: 14 Mast: 35	56 24 20
Her2	58 42	58 42
Cathepsin	93 7	95 5
HRT*	55 43	70 30
XRT	99 1	98 2
Subtypes*	LA: 17 LB: 16 HLB: 26 NLH: 31 3N: 10	14 14 26 32 14

FH (family history of cancer); multifocal (having more than one tumor in the breast); ER (estrogen receptor); PR (progesterone receptor); P53 (tumor protein 53); surgery: type of surgery (MRM: modified radical mastectomy, BCS: breast-conserving surgery, Mast: mastectomy); Her2 (epidermal growth factor receptor-2); Cathepsin (Cathepsin-D); HRT (hormone therapy); XRT (radiotherapy); subtypes: cancer molecular subtypes (LA: luminal A, LB: luminal B; HLB: HER2-positive luminal B, NLH: non-luminal Her2, 3N: triple negative). The characteristics were shown as positive and negative percentages (i.e. relative frequent table) for binary variables, respectively and the mode was underlined. The sample size was 579 and the recurrence prevalence was 19.3%.

*

Statistically significant features (P < 0.05).

Table 4: Comparison of the tumor "histological grade" feature (with ordinal measurement scales) of included subjects with/without cancer recurrence.

Grade categories	With recurrence	Without recurrence
1	22	10
2	44	59
3	33	29
4	1	2

The characteristics were shown as percentages (i.e. relative frequent table), and the mode (the most frequent item) was underlined in each category (recurrent or non-recurrent patient groups). Tumor grade was not statistically significant in subjects with/without recurrence. The sample size was 579 and the recurrence prevalence was 19.3%.

In our study, discretization was used on the following attribute intervals: age at diagnosis, NR and TS. In machine learning, discretization refers to the procedure of partitioning continuous attributes to discretized variables as to improve the classification performance [84,85]. We used a set of cut points taken from the literature, indicating the diagnosis properties of individual attributes as below to create categorical variables.

Age was categorized (Age.cat) as less than 40 years and more than 40 years [86]. NR was categorized (NR.cat) in ≤ 0.25 and > 0.25 [87,88]. TS was also categorized (TS.cat) as ≤ 2 (T1), (2,5) (T2) and > 5 (cm) (T3) [89–91].

The value of the fitness function (F-score on the training set) and the test set performance (F-score on the test set) at some iteration were shown in Fig. 2. The following categorical features were kept after running PSO and BDT: PR, HRT, Surgery, age.cat, TS.cat and NR.cat. Among the features with interval measurement scale, Ki67 and T. Node were not selected by BDT. In fact BDT uses entropy principle to choose the best discriminative features [92]. The estimated weight of the selected interval-scale feature I. Node was 0.74.

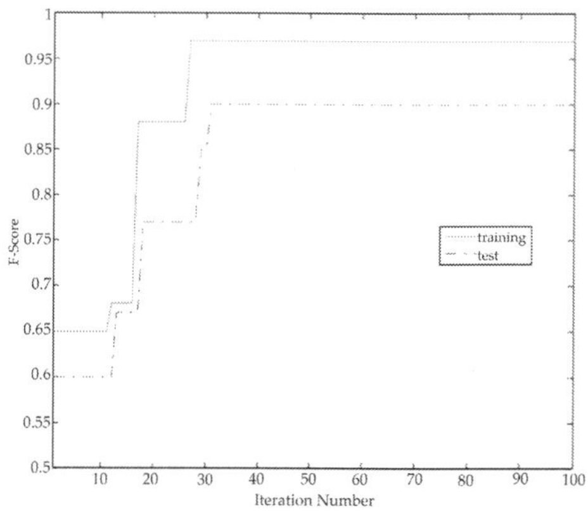

Figure 2: The value of the fitness function (F-Score of the proposed classifier (HPBCR) on the training set-solid line) and the F-Score on the test set (dash-dot line) during optimization procedure. The termination criterion was only the maximum number of iterations (i.e. 100) in this plot.

The performance of the classifiers on the test set (when trained on the training set) using the hold-out validation method was shown in Table 5. The performance of the classifiers was further assessed using 4-fold cross validation (Table 6). The confusion matrix of HPBCR on the test set was shown in Table 7. The performance of HPBCR was statistically significantly higher than that of three other classifiers (McNemar's test; $P < 0.05$). HPBCR showed limited discriminant power (DP = 1.28), very good diagnosis accuracy (AUC = 0.90), and excellent agreement with the gold standard (Kappa = 0.83). The minimum statistical power and maximum Type I error (α) were 77% and 0.07, respectively in HPBCR (Table 6). Overall, the average statistical power and Type I error of HPBCR was 80% and 0.04, respectively (Table 6). The training time of HPBCR, SVM, DT and MLP was 412.10 ± 14.50 (s), 66.20 ± 11.75 (s), 52.00 ± 4.50 (s) and 132.45 ± 15.50 (s), while the validation time was 0.62 ± 0.31 (s), 0.49 ± 0.24 (s), 0.39 ± 0.33 (s) and 0.49 ± 0.31 (s), respectively. The average running time was the average of 10 runs over 405 and 174 subjects in the training and test sets respectively (hold-out 70%) on an Intel Core-i7 2 GHz CPU with 4 GB of RAM.

Table 5: The holdout performance estimate (%) of the selected classifiers.

Method	Se %	Sp%	Acc%	Pr%	F-score %	Alpha	Beta	AUC	MCC	DOR	DP	Kappa
HPBCR	81	98	90	97	89	0.02	0.19	0.90	0.81	208.9	1.28	0.83
SVM	67	88	78	84	75	0.12	0.33	0.77	0.57	14.9	0.65	0.73
Decision tree	75	78	77	79	77	0.22	0.25	0.76	0.58	10.6	0.57	0.72
MLP	69	81	76	79	73	0.19	0.31	0.75	0.52	10.63	0.57	0.71

Se: sensitivity, Sp: specificity, Acc: accuracy, Pr: precision, AUC: area under the curve, MCC: Matthews correlation coefficient, DOR: diagnostics odds ratio; DP: discriminant power; SVM: supported vector machines; MLP: multilayer perceptron artificial neural network. Seventy percent of the data was used for training and the rest for validation.

Table 6: The performance of the classifiers in percent based on 4-fold cross-validation (mean ± SD) [min, max].

Method	Sensitivity	Specificity	Accuracy	Precision
Proposed method	80.0 ± 0.3[77.0, 80.5]	96.1 ± 1.0[93.0, 95.0]	89.2 ± 0.6[85.0, 89.0]	96.5 ± 1.0[95.0, 98.2]
SVM	73.0 ± 1.0[66.0, 74.0]	85.3 ± 1.2[80.0, 89.0]	77.6 ± 0.4[76.1, 78.9]	82.1 ± 0.3[80.0, 84.8]
Decision tree	74.0 ± 0.9[71.8, 76.0]	77.0 ± 0.8[75.1, 78.5]	77.1 ± 0.2[76.6, 77.5]	78.1 ± 0.6[77.3, 80.2]
MLP	67.5 ± 0.2[66.2, 67.9]	84.3 ± 1.0[80.8, 85.0]	76.0 ± .0.3[75.0, 76.8]	84.2 ± 1.8[79.9, 86.0]

SVM: supported vector machines; MLP: multilayer perceptron artificial neural network.

Table 7: The overall confusion matrix of HPBCR on the test set.

	Total population	**Condition (as determined by "gold standard")**	
		Condition positive	**Condition negative**
Test outcome	Test outcome positive	True positive 51	False positive (Type I error) 2
	Test outcome negative	False negative (Type II error) 11	True negative 110

Gold standard: recurrent cancer; test outcome: the decision of the proposed diagnosis system (HPBCR). The classifier was trained on the training set (70% of the samples) and its performance was shown on the test set.

DISCUSSION

The Risk Factors of the Cancer Recurrence

In our study, the following risk factors were selected by HPBCR to predict breast cancer recurrence: Age at diagnosis, TS, NR, I. Node, PR, HRT and type of surgery. Approximately 7% of women with breast cancer are diagnosed before the age of 40 years. It has been shown in the literature that younger age is not only an independent predictor of adverse outcome in survival rate [86], but it is also a risk factor in recurrence metastasis [89,93]. Bock et al. showed a hazard ratio of 2.8 for local recurrence in patients less than 35 years compared to those above 50 years [94]. Bharat et al. estimated the risk of breast cancer recurrence for women diagnosed below the age of 40 to be 1.53 times higher than in those diagnosed above 40 years [95]. In our study, the first and second age quartile for patients without recurrence was 41 and 47 years compared with 38 and 43 years for those with recurrence. Meanwhile, patients who were less than 40 years old had 33.5% higher hazard of recurrent metastasis in comparison with more than 40 year old patients [89]. In fact, the odds of recurrence in the young (< 40 years) and old (> 40 years) population in our study was 0.43 and 0.19, respectively. Such a trend is in agreement with the literature. Meanwhile, in recurrent cancer patients, the odds ratio (OR) associated with age ≤ 40 compared with age > 40 was 1.8, indicating that discretized age was a good feature for the developed prognosis system.

TS is an independent prognostic factor, with distant recurrence rates increasing with larger tumor size [96]. It has been in fact an important recurrence prognosis factor in several studies [97–100]. It has been shown in the literature that the median time to the development of metastatic disease shortens as tumor size increases [101–103]. In our study, the recurrence odds for TS ≤ 2 (T1), $2 < TS \leq 5$ (T2), and TS > 5 (T3) was 0.13, 0.27 and 0.29, respectively showing that increased tumor size is associated with higher probability of cancer recurrence.

NR was shown to be a prognosis factor for cancer recurrence in the literature [100]. Voordeckers et al. showed that NR is one of the main prognosis risk factors for cancer recurrence [104]. Elkhodary et al. showed that the hazard ratio of breast cancer recurrence risk increases when NR increases [105]. Tazhibi et al. indicated that NR (≤ 0.25 vs. > 0.25) is associated with hazard of distance recurrence [89]. In our study, the recurrence odds for the NR ≤ 0.25, and NR > 0.25 was 0.14, and 0.43, respectively showing that increased NR is associated with higher probability of cancer recurrence. Meanwhile, in recurrent cancer patients, the OR associated with NR > 0.25 compared with NR ≤ 0.25 was 1.6, indicating that discretized NR was a good feature for the developed prognosis system.

It was shown in the literature that I. Node is an important predictive factor for recurrence risk [106,107]. T. Node was also considered a predictor for cancer recurrence [108]. Although SFS selected this feature, it was not used in BDT. One explanation could be that in our dataset, I. Node and T. Node were moderately correlated ($r = 0.56$; $P < 0.001$). Also, NR defined as the ratio between I. Node to T. Node, was also used by the classifier. Thus, adding T. Node could not increase the amount of information in the prognosis system. In fact, reducing the number of correlated features improves the classification accuracy [109]. Higher Ki67 expression, on the other hand, was shown to be a prognosis indicator of increased cancer recurrence in the literature [110]. In our study, the second quartile of Ki67 was 14.3 and 17.4 in non-recurrent and recurrent cancer patient, respectably. Although SFS selected this feature, it was not used in BDT either. In our study, Higher Ki67 index tumors showed larger size (TS), and more lymph node involvement (I. Node), which is in agreement with what was reported in the literature [111]. This could be the justification that the final prognosis system did not use Ki67 expression.

PR expression was shown to be an independent recurrence prognostic variable in breast cancer in the literature [112–114]. Purdie et al. showed that absent PR expression was significantly associated with poor cancer recurrence prognosis [112]. In our study, the recurrence odds for the PR + and PR − was 0.20, and 0.31, respectively which is in agreement with what was mentioned in the literature. Meanwhile, in non-recurrent cancer patients, such odds associated with PR + compared with PR − was 1.6, indicating that PR was a distinctive feature for the developed prognosis system.

It was shown that HRT reduces the risk of breast cancer recurrence in the literature [115]. In our study, the recurrence odds for the HRT + and HRT − was 0.20, and 0.35, respectively which is in agreement with what was mentioned in the literature. Meanwhile, in non-recurrent cancer patients, the recurrence odds associated with HRT + compared with HRT − was 2.4, indicating that HRT was a suitable predictor for the developed prognosis system. It was proposed in the literature that the impact of type of surgery on the breast cancer recurrence might be dependent on age [116]. In our study, the non-recurrence odds for young and old patients underwent MRM surgery was 2.8 and 5.4, respectively. Such odds for the BCS surgery was 4.8 and 8.3. This could in principal support that the performance of the surgery is age dependent.

The Properties and Performance of HPBCR

In our study, among 1085 breast cancer patients, only 579 subjects with complete information and without primary metastasis were included in the prediction. Thus, subjects with missing data were excluded from the analysis. Missing data can introduce bias and reduce efficiency [117]. It is possible to use multiple imputation to handle such incomplete data. Although multiple imputation has potential to improve the validity of medical research, it requires the user to model the missing variable distribution. The validity of the imputed results highly depends on such modeling and underlying assumptions [118].

In our algorithm, SFS was first used to identify the discriminative features. Such a selection was then refined by PSO and also the importance of the interval features, represented by the estimated weights, was identified. Overall, the embedded FS was used in our study where FS is performed as part of the learning procedure. PSO is a meta-heuristics population-based stochastic optimization method. In principle, it is possible to use any other method in this category that could deal with continuous (feature weights) and discreet (feature selection) problems. We also tested genetic algorithm (GA) [119,120]. In multiple runs, both PSO and GA converged to an identical optimum solution. However, GA required about 150 iterations in comparison with 50 iterations in PSO (Fig. 2), for converging.

An alternative FS method is using multivariate FS namely as multiple logistic regression (MLR), taking into account the relationship between different features. MLR, known as feature vector machine in machine learning, can be used to select statistically significant features [84,121]. MLR was also tested in our study. However, due to the lack of fit of the MLR model, univariate SFS was used.

In HPBCR, F-score was used as the fitness function. When the dataset is imbalanced, for example in our case where the recurrence prevalence was 19.3, overall accuracy is not a suitable fitness measure. In fact, choosing the wrong objective function while developing a classification model can introduce bias towards majority [122]. Meanwhile HPBCR could be used for other mixed data where each instance in a data set is described by more than one type of measurement scale attribute [33]. In this case, different attributes must be treated differently. In our algorithm, different attributes were manipulated distinctly.

The average sensitivity and specificity of HPBCR are estimated as 80% and 96% (Table 6). Also, considering the maximum Types I ($\alpha = 0.07$) and II errors ($\beta = 0.23$), maximum false discovery rate (FDR = 5%), guarding against testing hypotheses suggested by the data (Type III errors [123]) done by cross-validation, very good diagnosis accuracy (AUC = 0.90), and finally the "excellent agreement" between the gold standard and the outcome of the system (Kappa = 0.83), HPBCR is promising for clinical prognosis tests. HPBCR significantly outperformed the other systems namely as MLP, DT and SVM (McNemar's test; $P < 0.05$). The other three tested classifiers showed poor DP, good diagnosis accuracy and good agreement with the gold standard (Table 5). The Types I and II errors of the proposed algorithm were better than those obtained by the other classifies (Tables 5 and 6). In HPBCR the average Type I error was less than 0.05 while the average statistical power was higher than 80% in the hold-out (Table 5) and cross validation (Table 6). However, the maximum Type I error (0.07) and minimum power (77%) in the folds used for cross validation were beyond the acceptable clinical ranges and must be improved. The variation of the performance indices in different folds was low (Table 6), showing consistency in the results of HPBCR.

The performance of HPBCR was further compared with the state-of-the-art. Kim et al. [28] used hold-out validation method (70%: training set; 30% test set) and reported the performance of the algorithm on the test set as the AUC of 0.85, sensitivity of 89%, specificity of 73%, precision of 75% (FDR = 25%) and accuracy of 85%. In a similar setting (Table 5), HPBCR had AUC of 0.90, sensitivity of 81%, specificity of 98%, precision of 97% (FDR = 3%) and accuracy of 90%. Our algorithm outperformed the other prediction method in terms of AUC, FDR and accuracy.

NPI method proposed by Galea et al. [27] was implemented and tested in our dataset. The value of the performance measures AUC, sensitivity, specificity, precision, and accuracy were 0.51, 100%, 2%, 25% and 26%, respectively. This method is not suitable in our study because of high Type I error ($\alpha = 0.98$).

Additional Analysis

The contribution of BDT and PSO to the overall classification performance was also considered. PSO was excluded from the algorithm flowchart and the resulting hold-out performance was reported in Table 8 (Scenario 1), compared with what obtained in the HPBCR (baseline). HPBCR significantly outperformed the modified prognosis system (McNemar's test; $P < 0.05$). Particularly, discriminant power is deteriorated when excluding PSO in the developed algorithm. Moreover, the modified system is not clinically reliable (Power = 78%, Type I error = 0.09). PSO is in fact used for embedded feature selection and is a necessary section of the program.

Table 8: The holdout performance estimate (%) of the prognosis system with modifications.

Scenario	Se%	Sp%	Acc%	Pr%	F-score %	Alpha	Beta	AUC	MCC	DOR	DP	Kappa
Baseline	81	98	90	97	89	0.02	0.19	0.90	0.81	208.9	1.28	0.83
1	78	81	78	93	86	0.18	0.21	0.79	0.51	15.1	0.65	0.78
2	81	98	90	96	89	0.02	0.19	0.90	0.81	208.9	1.28	0.83
3	80	96	88	97	87	0.04	0.2	0.88	0.79	116.9	1.14	0.81

Baseline: original developed HPBCR; Scenario 1: excluding PSO from the algorithm, 2: using SVM classifier instead of BDT, 3: using MLP classifier instead of BDT, Se: sensitivity, Sp: specificity, Acc: accuracy, Pr: precision, AUC: area under the curve, MCC: Matthews correlation coefficient, DOR: diagnostics odds ratio; DP: discriminant power; SVM: supported vector machines; MLP: multilayer perceptron artificial neural network. Seventy percent of the data was used for training and the rest for validation.

Moreover, SVM (Scenario 2) and MLP (Scenario 3) classifiers were used instead of BDT whose features were tuned using PSO. The performance of these two modified prognosis systems was comparable to what obtained in HPBCR (Table 8; McNemar's test; $P > 0.05$). Thus, BDT could be substituted by SVM and MLP. However, it was shown in the literature that BDT is as good as SVM and MLP [124]. Meanwhile, bagging avoids over-fitting. Using BDT, it was possible to implement HPBCR online efficiently (Section 2.7).

However, the performance of SVM or MLP when PSO is used for embedded feature selection (Table 8) was significantly better than that of SVM or MLP without using PSO (Table 5). This is the reason that hybrid feature selection performed by PSO is essential in the proposed prognosis system. Also, the training time of SVM + PSO, and MLP + PSO was 415.77 ± 10.50 (s) and 710.02 ± 22.55 (s), while the validation time was 0.50 ± 0.27, and 0.50 ± 0.33, respectively (hold-out 70%). Since at each iteration, feature sets are calculated for each 20 particle, the training time is more than cases where PSO is not used.

Final Considerations

This study was a validation of the prognostic performance of HPBCR. The limitation of the current study is that it was a retrospective, single-institution study. The results might have been biased on the basis of available specimen and patterns of referral to the hospital. The sample size must be increased as to improve the statistical power in our prognosis system [125]. The output of the classification system is not fuzzy. It will be useful to add risk assessment to the algorithm and report the risk of recurrence which is the focus of our future activity.

CONCLUSIONS

We proposed a computer-aided prognosis system to identify breast cancer recurrence within 5 years after diagnosis. This method is accurate and precise and could be possibly used by clinicians to make proper treatment plan.

ACKNOWLEDGEMENTS

This work was supported by the People Programme (Marie Curie Actions) of the European Union Seventh Framework Programme (FP7/2007–2013) under REA grant agreement no. 600388 (TECNIOspring Programme), from the Agency for Business Competitiveness of the Government of Catalonia, ACCIÓ and from Spanish Ministry of Economy and Competitiveness- Spain (project DPI2014-59049-R). This work was supported in part by the University of Isfahan and Isfahan University of Medical Sciences.

REFERENCES

1. Doi K. Computer-aided diagnosis in medical imaging: historical review, current status and future potential. Comput Med Imaging Graph 2007;31:198–211.
2. Castellino RA. Computer aided detection (CAD): an overview. Cancer Imaging 2005;5:17–9. http://dx.doi.org/10.1102/1470-7330.2005.0018.
3. Rangayyan RM, Ayres FJ, Desautels JL. A review of computer-aided diagnosis of breast cancer: toward the detection of subtle signs. J Franklin Inst 2007;344: 312–48.
4. Kim JH. Computer-aided diagnosis for lung cancer. J Lung Cancer 2004;3:67–70.
5. Oweis R, Hijazi L. A computer-aided ECG diagnostic tool. Comput Methods Programs Biomed 2006;81:279–84.
6. Katsumata A, Fujita H. Progress of computer-aided detection/ diagnosis (CAD) in dentistry CAD in dentistry. Jpn Dent Sci Rev 2014;50:63–8.

7. Kourou K, Exarchos TP, Exarchos KP, Karamouzis MV, Fotiadis DI. Machine learning applications in cancer prognosis and prediction. Comput Struct Biotechnol J 2015; 13:8–17.
8. Hadjiiski L, Sahiner B, Chan HP. Advances in computer-aided diagnosis for breast cancer. Curr Opin Obstet Gynecol 2006;18:64–70. http://dx.doi.org/10.1097/01. gco.0000192965.29449.da.
9. Anand P, Kunnumakara AB, Sundaram C, Harikumar KB, Tharakan ST, Lai OS, et al. Cancer is a preventable disease that requires major lifestyle changes. Pharm Res 2008;25:2097–116. http://dx.doi.org/10.1007/s11095-008-9661-9.
10. Kuper HBP, Adami HO. Tobacco use and cancer causation: association by tumour type. J Intern Med 2002;252:206–24.
11. Wild BWSaCP. World cancer report 2014. World Health Organization; 2014.
12. Siegel R, Ma J, Zou Z, Jemal A. Cancer statistics, 2014. Cancer J Clin 2014;64:9–29.
13. Torre LA, Bray F, Siegel RL, Ferlay J, Lortet-Tieulent J, Jemal A. Global cancer statistics, 2012. CA Cancer J Clin 2015;65:87–108. http://dx.doi.org/10.3322/caac.21262.
14. Thomsen AKJ. Chemoprevention of breast cancer. Am J Health Syst Pharm 2008.
15. Mousavi SM, Guoya MM, Ramazani R, Davanlou M, Hajsadeghi N, Seddighi Z. Cancer incidence and mortality in Iran. Oxf J Med Health Ann Oncol 2009;20: 556 –63.
16. Jazayeri SB, Saadat S, Ramezani R, Kaviani A. Incidence of primary breast cancer in Iran: ten-year national cancer registry data report. Cancer Epidemiol 2015;39: 519–27.
17. Maracy MR, Moradpour F, Hosseini SM, Tirani M. Cancer incidence and prevalence in Isfahan: application of mortality data to estimates and projects for the period 2001–2015. Int J Prev Med 2012;3:867–74.
18. Mokarian F, Ramezani MA, Heydari K, Tabatabaeian M, Tavazohi H. Epidemiology and trend of cancer in Isfahan 2005–2010. J Res Med Sci 2011;16:1228–33.

19. Dunnwald LK, Rossing MA, Li CI. Hormone receptor status, tumor characteristics, and prognosis: a prospective cohort of breast cancer patients. Breast Cancer Res 2007;9:R6. http://dx.doi.org/10.1186/bcr1639.
20. de Martel CFJ, Franceschi S, et al. Global burden of cancers attributable to infections in 2008: a review and synthetic analysis. Lancet Oncol 2012;13:607–15.
21. Lin R, Tripuraneni P. Radiation therapy in early-stage invasive breast cancer. Indian J Surg Oncol 2011;2:101–11. http://dx.doi.org/10.1007/s13193-011-0048-8.
22. Saphner T, Tormey DC, Gray R. Annual hazard rates of recurrence for breast cancer after primary therapy. J Clin Oncol 1996;14:2738–46.
23. Jemal A, Bray F, Center MM, Ferlay J, Ward E, Forman D. Global cancer statistics. CA Cancer J Clin 2011;61:69–90. http://dx.doi.org/10.3322/caac.20107.
24. Gonzalez-Angulo AMLJ, Broglio KR, Meric-Bernstam F, et al. High risk of recurrence for patients with breast cancer who have human epidermal growth factor receptor 2-positive, node-negative tumors 1 cm or smaller. J Clin Oncol 2009.
25. Weeks JC, Cook EF, O'Day SJ, Peterson LM, Wenger N, Reding D, et al. Relationship between cancer patients' predictions of prognosis and their treatment preferences. JAMA 1998;279:1709–14.
26. Zeng TLJ. Mixture classification model based on clinical markers for breast cancer prognosis. Artif Intell Med 2010;48:129–37.
27. Galea MH, Blamey RW, Elston CE, Ellis IO. The Nottingham prognostic index in primary breast cancer. Breast Cancer Res Treat 1992;22:207–19.
28. Kim W, Kim KS, Lee JE, Noh D-Y, Kim S-W, Jung YS, et al. Development of novel breast cancer recurrence prediction model using support vector machine. J Breast Cancer 2012;15:230–8.
29. Ahmad L, Eshlaghy A, Poorebrahimi A, Ebrahimi M, Razavi A. Using three machine learning techniques for predicting breast cancer recurrence. J Health Med Inform 2013;4:124.
30. Jafari-Koshki T, Mansourian M, Mokarian F. Exploring factors

related to metastasis free survival in breast cancer patients using Bayesian cure models. Asian Pac J Cancer Prev 2014;15:9673–8.

31. Wishart GC, Bajdik CD, Dicks E, Provenzano E, Schmidt MK, Sherman M, et al. PREDICT Plus: development and validation of a prognostic model for early breast cancer that includes HER2. Br J Cancer 2012;107:800–7. http://dx.doi.org/10. 1038/bjc.2012.338.
32. Shim HJ, Kim SH, Kang BJ, Choi BG, Kim HS, Cha ES, et al. Breast cancer recurrence according to molecular subtype. Asian Pac J Cancer Prev 2014;15:5539–44.
33. Marateb HR, Mansourian M, Adibi P, Farina D. Manipulating measurement scales in medical statistical analysis and data mining: a review of methodologies. J Res Med Sci 2014;19:47–56.
34. Guyon I, Elisseeff A. An introduction to variable and feature selection. J Mach Learn Res 2003;3:1157–82.
35. Saeys Y, Inza I, Larranaga P. A review of feature selection techniques in bioinformatics. Bioinformatics 2007;23:2507–17. http://dx.doi. org/10.1093/bioinformatics/ btm344.
36. Nilsson R. Statistical feature selection: with applications in life science. SE58183 Linköping, Sweden: Linköping University, The Institute of Technology; 2007, 191 http://www.diva-portal.org/ smash/get/diva2:17577/FULLTEXT01.pdf.
37. Destercke S, Strauss O. Kolmogorov–Smirnov test for interval data. Springer; 2014
38. 416–25.
39. Hilton JF. The appropriateness of the Wilcoxon test in ordinal data. Stat Med 1996; 15:631–45.
40. Quinlan JR. Induction of decision trees. Mach Learn 1986;1:81–106.
41. Quinlan JR. C4. 5: programs for machine learning. Elsevier; 2014.
42. Salzberg SL. C4. 5: programs for machine learning by J. Ross Quinlan. Morgan Kaufmann Publishers, Inc., 1993. Mach Learn 1994;16:235–40.
43. Jerez-Aragonés JM, Gómez-Ruiz JA, Ramos-Jiménez G, Muñoz-Pérez J, Alba-Conejo E. A combined neural network and decision

trees model for prognosis of breast cancer relapse. Artif Intell Med 2003;27:45–63.

44. Markey MK, Tourassi GD, Floyd CE. Decision tree classification of proteins identified by mass spectrometry of blood serum samples from people with and without lung cancer. Proteomics 2003;3:1678–9.
45. Tan AC, Gilbert D. Ensemble machine learning on gene expression data for cancer classification; 2003.
46. Podgorelec V, Kokol P, Stiglic B, Rozman I. Decision trees: an overview and their use in medicine. J Med Syst 2002;26:445–63.
47. Bae J-M. The clinical decision analysis using decision tree. Epidemiol Health 2014; 36.
48. Breiman L. Bagging predictors. Mach Learn 1996;24:123–40.
49. Machová K, Barcak F, Bednár P. A bagging method using decision trees in the role of base classifiers. Acta Polytech Hung 2006;3:121–32.
50. Lavanya D, Rani KU. Ensemble decision tree classifier for breast cancer data. Int J Inf Technol Converg Serv 2012;2:17.
51. Valentini G, Muselli M, Ruffino F. Bagged ensembles of support vector machines for
52. gene expression data analysis. IEEE; 2003 1844–9.
53. Qu Y, Adam B-L, Yasui Y, Ward MD, Cazares LH, Schellhammer PF, et al. Boosted decision tree analysis of surface-enhanced laser desorption/ionization mass spectral serum profiles discriminates prostate cancer from noncancer patients. Clin Chem 2002;48:1835–43.
54. Eberhart RC, Kennedy J. A new optimizer using particle swarm theory. Proceedings of the 6th IEEE International Symposium on Micro Machine and Human Science (MHS '95); October 1995. p. 39–43. http://dx.doi.org/10.1109/MHS.1995.494215 Nagoya, Japan.
55. Sha D, Lin H-H. A multi-objective PSO for job-shop scheduling problems. Expert
56. Syst Appl 2010;37:1065–70.
57. Gopalakrishnan K. Particle swarm optimization in civil infrastructure

systems: state-of-the-art review. Metaheuristic applications in structures and infrastructures. London (UK): Elsevier; 2013. p. 49–76.

58. Durán O, Pérez L, Batocchio A. Optimization of modular structures using particle swarm optimization. Expert Syst Appl 2012;39:3507–15.
59. Marateb HR, McGill KC. Resolving superimposed MUAPs using particle swarm optimization. IEEE Trans Biomed Eng 2009;56:916.
60. Modiri A, Kiasaleh K. Permittivity estimation for breast cancer detection using particle swarm optimization algorithm. IEEE; 2011 1359–62.
61. Gopalakrishnan K, Ceylan H, Attoh-Okine NO. Intelligent and soft computing in infrastructure systems engineering: recent advances. Springer; 2009.
62. Wasserman PD, Schwartz T. Neural networks. II. What are they and why is everybody so interested in them now? IEEE Expert 1988;3:10–5.
63. Isa I, Saad Z, Omar S, Osman M, Ahmad K, Sakim HM. Suitable MLP network activation functions for breast cancer and thyroid disease detection. IEEE; 2010 39–44.
64. Cortes C, Vapnik V. Support-vector networks. Mach Learn 1995;20:273–97.
65. Akay MF. Support vector machines combined with feature selection for breast cancer diagnosis. Expert Syst Appl 2009;36:3240–7.
66. Widodo A, Yang B-S. Support vector machine in machine condition monitoring and fault diagnosis. Mech Syst Signal Process 2007;21:2560–74.
67. George, G, Raj, VC. Review on feature selection techniques and the impact of SVM for cancer classification using gene expression profile. (arXiv preprint arXiv:1109. 1062 2011).
68. Wu K-P, Wang S-D. Choosing the kernel parameters for support vector machines by the inter-cluster distance in the feature space. Pattern Recognit 2009;42:710–7.
69. Sammut C, Webb GI. Encyclopedia of machine learning. Springer Science & Business Media; 2011.

70. Webb AR. Statistical pattern recognition. John Wiley & Sons; 2011.
71. Metz CE. Basic principles of ROC analysis. Elsevier; 1978 283–98.
72. Lalkhen AG, McCluskey A. Clinical tests: sensitivity and specificity. Contin Educ Anaesth Crit Care Pain 2008;8:221–3.
73. [70] Huang J, Ling CX. Using AUC and accuracy in evaluating learning algorithms. IEEE Trans Knowl Data Eng 2005;17:299–310.
74. Glas AS, Lijmer JG, Prins MH, Bonsel GJ, Bossuyt PM. The diagnostic odds ratio: a
75. single indicator of test performance. J Clin Epidemiol 2003;56:1129–35.
76. Sokolova M, Japkowicz N, Szpakowicz S. Beyond accuracy, f-score and ROC: a family of discriminant measures for performance evaluation. Hobart, Australia: Springer-Verlag; 2006 1015–21.
77. Mert A, Kilic N, Bilgili E, Akan A. Breast cancer detection with reduced feature set. Comput Math Methods Med 2015;2015, 265138. http://dx.doi.org/10.1155/2015/ 265138.
78. Simundic A, Measures of diagnostic accuracy: basic definitions. eJIFCC 2008; 19 (4). 2013.
79. Smeeton NC. Early history of the kappa statistic. JSTOR 1985.
80. Fleiss JL, Levin B, Paik MC. Statistical methods for rates and proportions. John Wiley
81. & Sons; 2013.
82. Banerjee A, Chitnis UB, Jadhav SL, Bhawalkar JS, Chaudhury S. Hypothesis testing, type I and type II errors. Ind Psychiatry J 2009;18:127–31. http://dx.doi.org/10. 4103/0972-6748.62274.
83. Ellis PD. The essential guide to effect sizes: statistical power, meta-analysis, and the interpretation of research results. Cambridge; New York: Cambridge University Press; 2010.
84. Colquhoun D. An investigation of the false discovery rate and the misinterpretation of p-values. Open Sci 2014;1:140216.
85. Guilherme HMO, Nesbitt GC, Murphy JG, Habermann TM. Mayo Clinic medical manual and Mayo Clinic internal medicine review. 7th ed. MN, USA: CRC Press Rochester; 2007.

86. Dietterich TG. Approximate statistical tests for comparing supervised classification learning algorithms. Neural Comput 1998;10:1895–923.
87. Eliasziw M, Donner A. Application of the McNemar test to non-independent matched pair data. Stat Med 1991;10:1981–91.
88. Demšar J. Statistical comparisons of classifiers over multiple data sets. J Mach Learn Res 2006;7:1–30.
89. Marateb HR, Goudarzi S. A noninvasive method for coronary artery diseases diagnosis using a clinically-interpretable fuzzy rule-based system. J Res Med Sci 2015;20:214–23.
90. Huan L, Setiono R. Feature selection via discretization. IEEE Trans Knowl Data Eng 1997;9:642–5. http://dx.doi.org/10.1109/69.617056.
91. Anders CK, Johnson R, Litton J, Phillips M, Bleyer A. Breast cancer before age 40 years. Semin Oncol 2009;36:237–49. http://dx.doi.org/10.1053/j.seminoncol. 2009.03.001.
92. Keam B, Im SA, Kim HJ, Oh DY, Kim JH, Lee SH, et al. Clinical significance of axillary nodal ratio in stage II/III breast cancer treated with neoadjuvant chemotherapy. Breast Cancer Res Treat 2009;116:153–60. http://dx.doi.org/10.1007/s10549-008- 0160-9.
93. Truong PT, Berthelet E, Lee J, Kader HA, Olivotto IA. The prognostic significance of the percentage of positive/dissected axillary lymph nodes in breast cancer recurrence and survival in patients with one to three positive axillary lymph nodes. Cancer 2005;103:2006–14. http://dx.doi.org/10.1002/cncr.20969.
94. Tazhibi M, Fayaz M, Mokarian F. Detection of prognostic factors in metastatic breast cancer. J Res Med Sci 2013;18:283–90.
95. Narod SA. Tumour size predicts long-term survival among women with lymph node-positive breast cancer. Curr Oncol 2012;19:5. http://dx.doi.org/10.3747/co. 19.1043.
96. Saadatmand S, Bretveld R, Siesling S, Tilanus-Linthorst MMA. Influence of tumour stage at breast cancer detection on survival in modern times: population based study in 173,797 patients. BMJ Br Med J 2015;351.

97. Zadrozny B, Elkan C. Obtaining calibrated probability estimates from decision trees and naive Bayesian classifiers. Citeseer; 2001 609–16.
98. Assi HA, Khoury KE, Dbouk H, Khalil LE, Mouhieddine TH, El Saghir NS. Epidemiology and prognosis of breast cancer in young women. J Thorac Dis 2013;5(Suppl. 1):S2–8. http://dx.doi.org/10.3978/j.issn.2072-1439.2013.05.24.
99. de Bock GH, van der Hage JA, Putter H, Bonnema J, Bartelink H, van de Velde CJ. Isolated loco-regional recurrence of breast cancer is more common in young patients and following breast conserving therapy: long-term results of European Organisation for Research and Treatment of Cancer studies. Eur J Cancer 2006; 42:351–6. http://dx.doi.org/10.1016/j.ejca.2005.10.006.
100. Bharat A, Aft RL, Gao F, Margenthaler JA. Patient and tumor characteristics associated with increased mortality in young women (b or =40 years) with breast cancer. J Surg Oncol 2009;100:248–51. http://dx.doi.org/10.1002/jso.21268.
101. Cianfrocca M, Goldstein LJ. Prognostic and predictive factors in early-stage breast cancer. Oncologist 2004;9:606–16. http://dx.doi.org/10.1634/theoncologist.9-6-
102. 606.
103. Voogd AC, Nielsen M, Peterse JL, Blichert-Toft M, Bartelink H, Overgaard M, et al. Differences in risk factors for local and distant recurrence after breast-conserving therapy or mastectomy for stage I and II breast cancer: pooled results of two large European randomized trials. J Clin Oncol Off J Am Soc Clin Oncol 2001;19: 1688–97.
104. Arriagada R, Rutqvist LE, Johansson H, Kramar A, Rotstein S. Predicting distant dissemination in patients with early breast cancer. Acta Oncol 2008;47:1113–21. http://dx.doi.org/10.1080/02841860701829661.
105. Silvestrini R, Daidone MG, Di Fronzo G, Morabito A, Valagussa P, Bonadonna G. Prognostic implication of labeling index versus estrogen receptors and tumor size in node-negative breast cancer. Breast Cancer Res Treat 1986;7:161–9.

106. O'Reilly SM, Camplejohn RS, Barnes DM, Millis RR, Rubens RD, Richards MA.
107. Node-negative breast cancer: prognostic subgroups defined by tumor size and flow cytometry. J Clin Oncol Off J Am Soc Clin Oncol 1990;8:2040–6.
108. Carter CL, Allen C, Henson DE. Relation of tumor size, lymph node status, and survival in 24,740 breast cancer cases. Cancer 1989;63:181–7.
109. Rosen PP, Groshen S, Kinne DW, Norton L. Factors influencing prognosis in
110. node-negative breast carcinoma: analysis of 767 T1N0M0/T2N0M0 patients with long-term follow-up. J Clin Oncol Off J Am Soc Clin Oncol 1993;11:2090–100.
111. [103] Fisher B, Slack NH, Bross ID. Cancer of the breast: size of neoplasm and prognosis. Cancer 1969;24:1071–80.
112. Voordeckers M, Vinh-Hung V, Van de Steene J, Lamote J, Storme G. The lymph node ratio as prognostic factor in node-positive breast cancer. Radiother Oncol 2004;70: 225–30. http://dx.doi.org/10.1016/j.radonc.2003.10.015.
113. Elkhodary TR, Ebrahim MA, Hatata EE, Niazy NA. Prognostic value of lymph node ratio in node-positive breast cancer in Egyptian patients. J Egypt Natl Canc Inst 2014;26:31–5. http://dx.doi.org/10.1016/j.jnci.2013.10.001.
114. Martin C, Cutuli B, Velten M. Predictive model of axillary lymph node involvement
115. in women with small invasive breast carcinoma: axillary metastases in breast carcinoma. Cancer 2002;94:314–22. http://dx.doi.org/10.1002/cncr.10229.
116. Lin YC, Chen SC, Chang HK, Hsueh S, Tsai CS, Lo YF, et al. Identifying good prognosis group of breast cancer patients with 1–3 positive axillary nodes for adjuvant cyclophosphamide, methotrexate and 5-fluorouracil (CMF) chemotherapy. Jpn J Clin Oncol 2005;35:514–9. http://dx.doi.org/10.1093/jjco/hyi143.
117. Weir L, Speers C, D'Yachkova Y, Olivotto IA. Prognostic significance of the number of axillary lymph nodes removed in

patients with node-negative breast cancer. J Clin Oncol Off J Am Soc Clin Oncol 2002;20:1793–9.

118. Li S, Oh S. Improving feature selection performance using pairwise pre-evaluation. BMC Bioinformatics 2016;17:1–13. http://dx.doi.org/10.1186/s12859-016-1178-3.
119. Yang XQ, Wang FB, Chen C, Peng CW, Zhang JF, Li Y. High Ki-67 expression is a poor prognostic indicator of 5-year recurrence free survival in patients with invasive breast cancer. Asian Pac J Cancer Prev 2011;12:3101–5.
120. Shokouh TZ, Ezatollah A, Barand P. Interrelationships between Ki67, HER2/neu, p53, ER, and PR status and their associations with tumor grade and lymph node involvement in breast carcinoma subtypes: retrospective-observational analytical study. Medicine 2015;94, e1359. http://dx.doi.org/10.1097/MD.0000000000001359.
121. Purdie CA, Quinlan P, Jordan LB, Ashfield A, Ogston S, Dewar JA, et al. Progesterone receptor expression is an independent prognostic variable in early breast cancer: a population-based study. Br J Cancer 2014;110:565–72. http://dx.doi.org/10.1038/ bjc.2013.756.

8 THE EFFECTS OF SHARED INFORMATION ON SEMANTIC CALCULATIONS IN THE GENE ONTOLOGY

Paul W. Bible[a], Hong-Wei Sun[b], Maria I. Morasso[c], Rasiah Loganantharaj[d], Lai Wei [a],

[A] State Key Laboratory of Ophthalmology, Zhongshan Ophthalmic Center, Sun Yat-sen University, Guangzhou 510060, China

[B] Biodata Mining and Discovery Section, Office of Science and Technology, Intramural Research Program, National Institute of Arthritis and Musculoskeletal and Skin Diseases,

Bethesda, Maryland

[C] Laboratory of Skin Biology, Intramural Research Program, National Institute of Arthritis and Musculoskeletal and Skin Diseases, Bethesda, Maryland

[D] Laboratory of Bioinformatics, Center for Advanced Computer Studies, University of Louisiana at Lafayette, Lafayette, Louisiana

ABSTRACT

The structured vocabulary that describes gene function, the gene ontology (GO), serves as a powerful tool in biological research. One application of GO in computational biology calculates semantic similarity between two concepts to make inferences about the functional similarity of genes. A class of term similarity algorithms explicitly calculates the shared information (SI) between concepts then substitutes this calculation into traditional term similarity measures such as Resnik, Lin, and Jiang-Conrath. Alternative SI approaches, when combined with ontology choice and term similarity type, lead to many gene-to-gene similarity measures. No thorough investigation has been made into the behavior, complexity, and performance of semantic methods derived from distinct SI approaches. We apply bootstrapping to compare the generalized performance of 57 gene-to-gene semantic measures across six benchmarks. Considering the number of measures, we additionally evaluate whether these methods can be leveraged through ensemble machine learning to improve prediction performance. Results showed that the choice of ontology type most strongly influenced performance across all evaluations. Combining measures into an ensemble classifier reduces cross-validation error beyond any individual measure for protein interaction prediction. This improvement resulted from information gained through the combination of ontology types as ensemble methods within each GO type offered no improvement. These results demonstrate that multiple SI measures can be leveraged for machine learning tasks such as automated gene function prediction by incorporating methods from across the ontologies. To facilitate future research in this area, we developed the GO Graph Tool Kit (GGTK), an open source C++ library with Python interface (github.com/paulbible/ggtk).

Keywords: Semantic similarity, Gene ontology, Function prediction, Machine learning, Protein–protein interaction, Gene expression

INTRODUCTION

Researchers developed the gene ontology (GO) to provide a structured vocabulary that consistently describes the characteristics of genes and proteins across different organisms [1,2]. Specific GO terms in this vocabulary annotate proteins by specifying the biological processes in which they participate, their enzymatic and molecular functions, and their location within the cell. As a structured vocabulary, GO explicitly defines the relationships between terms using a directed acyclic graph (DAG). These relationships serve to clarify terminology, for example by identifying when one term may be a more specialized from of another. Three separate ontologies exist that provide a DAG of terms and relationships used to describe biological processes (BP), molecular functions (MF), and cellular components (CC). These term structures are not fixed. The Gene Ontology Consortium makes frequent updates to GO modifying the relationship structure and adding or removing terms to better reflect the current understanding of biological functions.

The annotation of gene products with GO terms provides a valuable resource allowing the comparison of functions both within and between separate organisms. For each annotation of a term to a protein, GO provides evidence codes that allow researchers to consider the methods that produced each annotation. The use of GO plays an important role in the analysis of high-throughput experiments thanks in part to computational methods that utilize the rich domain knowledge encoded in GO annotations. The rigid, well-defined structure of GO proves to be an advantage that facilitates its integration into statistical and computational analyses. Methods of semantic similarity take advantage of this structure to quantify the similarity between the meaning of one term and another. Through semantic measures, the concept level knowledge stored in functional annotations provides the ability to quantify functional similarity between genes. Researchers have employed semantic methods for predicting protein protein interactions [3–6], prioritizing host-pathogen interactions [7], and automated function prediction [8,9]. Increasingly, researchers apply computational methods to infer new GO annotations. These annotations receive the evidence code *inferred from electronic annotation* (IEA), and recent studies show that these predicted annotations are increasingly reliable [7,10].

Since the seminal works by Lord et al. [11,12], semantic similarity applications have become established tools in computational biology and bioinformatics. Many diverse methods exist for calculating semantic similarity between terms. The reviews of Refs. [13], [14], and [15] provide a thorough overview of semantic measures used in the Gene Ontology and other biomedical ontologies. The review by Pesquita et al. [13]categorizes semantic similarity at the term level into edge-based and node-based methods. Edge-based methods usually quantify semantic similarity using a function of the paths between two terms in the graph. Node-based methods use properties derived from the terms and often include operations on the shared ancestors or descendants of two terms. Lord et al. adapted three well studied information theoretic semantic measures, Resnik [16], Lin [17], and Jiang-Conrath [18], using a corpus-based calculation of term probability and information content (IC). These methods calculate the similarity between two concepts by operating on their shared and unique information. Lord et al. calculated shared information using the IC of the most informative common ancestor (MICA) between two terms. Information content based semantic methods have been extensively studied and research suggests that IC offers a superior conception of a term's specificity over methods based on graph depth [14,19].

Couto et al. [20] developed an alternative approach to shared information. While the MICA shared information considers only a single ancestor, the alternative approach, called the graph-based similarity measure (GraSM), considers multiple inheritance of ontology terms using a path counting method. GraSM calculates the mean IC of disjunctive common ancestors between two concepts. Couto et al. recognized the modularity of using alternative shared information methods and substituted GraSM shared information into Resnik, Lin, and Jiang-Conrath deriving three new term similarity algorithms. This method was shown to improve the performance of semantic similarity on accepted evaluation metrics such as correlation with sequence and domain similarity. The computational cost of path counting poses a significant challenge to the real-time calculation of GraSM. Zhang and Lai [21] proposed a faster GraSM alternative called exclusively inherited shared information that calculates a subset of the disjunctive common ancestors. The modular separation of shared information from term similarity leads to some interesting properties. Any new conception of shared information immediately implies the construction of three new term similarity measures. The success of GraSM and alternative conceptions of shared information clarifies the need for a thorough exploration of shared information in the biomedical ontologies.

Extending *term* similarity to *gene* similarity requires methods operating on term sets. For two genes, represented as term sets, the all-pairs term similarity is calculated then summarized by aggregation methods such as the *min* [22], *max* [23], *average* [11], or *best-match average* (BMA) [24]. See Ref. [13] for a detailed description of aggregation methods.

Furthermore, genes are described by annotations from each of the independent ontologies (BP, MF, CC). The choices for shared information calculation, term similarity algorithm, aggregation method, and ontology type lead to a combinatorial increase in the number of gene similarity measures. Considering the number of measure that can be constructed, we address the following questions in this work. Are the methods derived from alternative shared information calculations truly distinct or do they offer no statical difference in practical applications? If these methods are not distinct, it would imply that computationally intensive algorithms, such as GraSM, can be replaced by more efficient alternatives. Some gene similarity methods avoid explicit calculation of shared information, term similarity, and aggregation. Methods based on the simple yet powerful Jaccard set similarity, such as SimUI [25], SimGIC [19], SimDIC [26], and SimUIC [26], have been shown to perform well on a variety of tasks and are computationally easy to compute. Can methods derived from modular combination with shared information outperform these more efficient methods? If the shared information-based methods are truly distinct, can the large quantity of measures be leveraged for performance gains through ensemble integration techniques?

This work presents extensive and novel research on the under-studied effects of shared information in semantic calculations in the gene ontology through robust evaluations of their performance on traditional and real-world tasks. As no method can determine *true* semantic similarity between genes [13], various methods for evaluation have been put forward to evaluate the performance of semantic similarity methods. Commonly used evaluations include correlation with sequence similarity [11,19], correlation with domain set similarity [13,20,27], correlations with gene expression [4,23,28], clustering genes into known pathways [3,5,21], and prediction of protein–protein interactions [3,6,29]. Based on these past works, we have constructed a new suite of six benchmarks to evaluate semantic similarity that provides evaluations for a broad range of use cases. These benchmarks are provided as lists of protein pairs and scores to facilitate use by other researchers. It is known that the GO annotations are incomplete and can suffer from bias and noise [30]. To address these issues, our evaluations use bootstrapping to provide robust statistical performance comparisons for each measure. Due to the large number of measures and the number of evaluations needed, no current tools addressed our simultaneous needs of speed and modularity. To achieve these goals, we have developed a new set of efficient, modular tools for working with GO graphs in C++, called the GO Graph Tool Kit (GGTK). With the aim of facilitating further research in the community, we provide an easy-to-use Python package that wraps the functionality of GGTK and release all GGTK code under the permissive BOOST License. GGTK will remain an ongoing open source project available at github.com/paulbible/ggtk.

METHODS

Calculating information content and shared information

Understanding the structure of the GO DAG provides insight into how semantic similarity is derived at the term level. In GO graphs, each vertex or node represents a term or concept and each edge presents a relationship such as *is_a* or *part_of* [1]. A root node in GO represents the most general concept (e.g. *biological_process*), and all other concepts are considered descendants of this term. Information theoretic semantic measures rely on assigning a value of probability to a term. Lord et al. [11]proposed defining a term's probability as the number of times it occurs in a corpus of annotations divided by the number of occurrences for all terms. A term occurs if it or *any of its descendants* appear in the corpus. Eq. (1) shows the definition of probability where o_t is the number of occurrences for a term t. The information content (IC) follows, in Eq. (2), as the negative log of probability. As o_t includes all appearances of child terms, $P(root) = 1$ and any non-root term t satisfies $P(t) < P(root)$. These definitions ensure IC is monotonically increasing toward more specific terms.

$$P(t) = \frac{o_t}{o_{root}} \qquad (1)$$

$$IC(t) = -\log(P(t)) \qquad (2)$$

Using the above definitions Lord et al. used the IC calculation in adapting Resnik [16], Lin [17], and Jiang-Conrath [18] to GO. The original shared information approach calculates the set of common ancestors (CA) between two ontology terms (Eq. (3)), and finds the most informative common ancestor (MICA) or equivalently, after Resnik [16], the probability of the minimum subsumer (Eq. (4)). The shared information of these traditional term similarity measures is defined in Eq. (5).

$$CA(t_1,t_2)=Ancestors(t_1)\cap Ancestors(t_2) \quad (3)$$

$$P_{ms}(t_1,t_2) = \min_{t \in CA(t_1,t_2)} \{P(t)\} \quad (4)$$

$$\begin{aligned} SI_{MICA}(t_1,t_2) &= IC\,(MICA(t_1,t_2)) \\ &= -\log(P_{ms}(t_1,t_2)) \end{aligned} \quad (5)$$

Using the above definitions, Resnik, Lin, and Jiang-Conrath semantic similarity (SS) can be re-written equivalently in terms of shared information (Eqs. (6), (7), and (8) respectively).

$$SS_{Resnik}(t_1,t_2) = SI(t_1,t_2) \quad (6)$$

$$SS_{Lin}(t_1,t_2) = \frac{2 * SI(t_1,t_2)}{IC(t_1) + IC(t_2)} \quad (7)$$

$$SS_{JC}(t_1,t_2) = 1 - (IC(t_1) + IC(t_2) - 2 * SI(t_1,t_2)) \quad (8)$$

Implementation considerations affecting time complexity

Before describing the different shared information algorithms and their time complexity, we will explain the implementation of GO used within the Go Graph Tool Kit (GGTK). Varying reports on the time complexity of semantic similarity algorithms have been put forward due to implicit assumptions about the representation of the GO DAG and associated annotations. Here we clarify the graph representation and its implementation. GGTK reads the ontology files (go-basic.obo) provided by geneontology.org [1], and stores the graph structure for the three disjoint ontologies (BP, MF, CC) in memory as adjacency lists. Let n be the number of terms in the ontology. Probability and information content are calculated from annotations taken from UniProt-GOA [31]. GGTK reads GO annotations from file, indexes annotations by gene, and provides gene-to-annotation look-ups in $O(1)$ time. Calculation of IC is performed after Lord et al. [11] and is accomplished in $O(n)$ time. Specifically, a depth first traversal visits every node in the GO graph. The number of annotation occurrences is calculated for each term starting with leaf nodes, moving to more general terms in the graph, and finishing with the root node. The number of cumulative occurrences for the root node is used to calculate term probability and IC after Eqs. (1) and (2) respectively. After the initial calculation, a function that maps a term to its IC is available as a map in which lookups are performed in $O(1)$ time. In this work, IC was calculated using human annotations (GOA Dec. 2016) including electronically inferred annotations and considering only *is_a* and *part_of* relationships.

The semantic similarity methods under study in this work all rely on the calculation of the set of common ancestors shared between two concepts. The DAG structure of GO complicates estimates of the size of this set with respect to the number of terms, n. For a tree based ontology, the size of the common ancestor sets is bounded by $O(\log n)$, but this complexity is not guaranteed for general DAGs. Using GGTK we compiled historical data on the average branching factor and number of ancestors of each term from 2006 to 2016. Fig. 1 shows summary data for the three ontologies BP, MF, and CC (considering only *is_a* and *part_of* relationships) as the graph topology evolved over 10 years. The number of nodes, mean branching factor (node degree), and mean ancestor number are growing over time. Fig. 1D shows that $\log n$ is an under estimate in the case of BP and CC but over estimates the average number of ancestors in MF. After Couto et al. [20], we will refer to the number of common ancestors as k in our complexity analysis where $|CA| = O(k)$. Based on our empirical analysis of the GO graph over the past 10 years, it appears that $O(k) \approx O(\log n)$ for current and past graphs; however, this relation may not hold in the future.

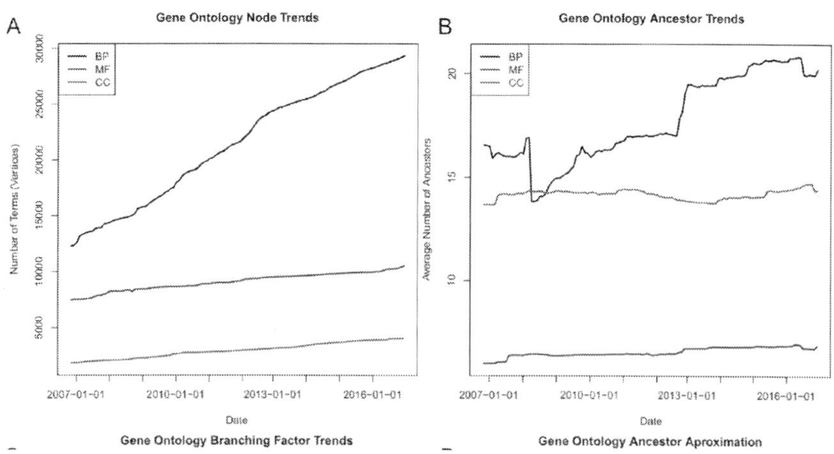

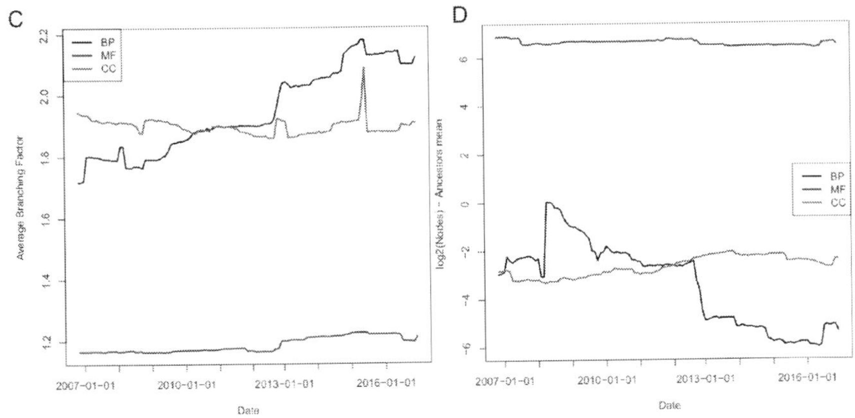

Figure 1: Changes in GO graph structure (using *is_a* and *part_of* relationships) over time lead to variations in (A) number of nodes, (B) mean ancestor number, and (C) the mean branching factor for each term. Panel D shows the log of the number of terms minus the mean ancestor number.

Shared information algorithms and their time complexity

Determining the shared information between two ontology terms is an area of ongoing research. In this work we address the effects of five unique shared information algorithms on their derived gene functional similarity measures. A thorough analysis of these methods must address their time complexity as well as qualitative features. In this section, we provide a brief description of the shared information algorithms analyzed in this work, describe their features, and address their time complexity. A more detailed description of the algorithms can be found in Appendix A. The source code for all methods is available at github.com/paulbible/ggtk. The following complexity bounds refer to the calculation of single-term-to-single-term shared information.

Common ancestor shared information

For purposes of evaluation, we developed a naive baseline algorithm called common ancestor shared information (CASI). CASI simply calculates the set of common ancestors and returns the mean IC as the shared information. This naive baseline serves as a useful tool for evaluating other shared information algorithms. The time complexity of

CASI is $O(k)$. Calculating the common ancestor set is performed in $O(k)$ and using GGTK's $O(1)$ IC map allows the mean to be calculated in $O(k)$.

Most informative common ancestor shared information

The most informative common ancestor (MICA) shared information remains the most common shared information measure as it forms the basis of traditional term similarity methods. (Eq. (5), above). MICA shared information is equivalent to Resnik term similarity [11,16], and is defined as the IC of the MICA. Due to its simplicity, it is one of the most widely used shared information measures and forms the foundation for many more sophisticated semantic algorithms. The time complexity of MICA is $O(k)$. Construction of the common ancestor set takes $O(k)$ and finding the term with maximum IC in the set is also $O(k)$.

Couto et al. 2007, GraSM

Couto et al. devised GraSM [20] as an alternative to MICA. They observed that different paths in the ontology represent different interpretations of concepts. They reasoned that ancestors with multiple interpretations, designated disjunctive common ancestors (DCA), should factor into the shared information. Calculation of the DCA involves determining if a common ancestor has unique paths to each of the two terms that are separate from the other ancestors under consideration. For each common ancestor, t_a, paths are counted to determine if t_a is disjoint from another ancestor, say t_b. If the number paths from t_a to the two input terms under consideration is greater than or equal to the number of paths from t_b to the input terms, t_a must be disjoint and represent some unique interpretation of the shared information by virtue of having a unique path to the terms that does not pass though t_b. This process is repeated for all pairs of common ancestors. A detailed description of the GraSM algorithm is provided in Appendix A.3. Couto et al. reported the GraSM time complexity as $O(k^2)$, but their implementation uses a pre-computed path number map. The term-to-term path number map can be calculated for a one-time cost of $O(n^2)$, where n is number of terms in the ontology. The parallelized version of this calculation can be performed by n separate topological sorts each taking $O(n)$ time. This 'one-time' cost could be quite large as $n \gg k$ and would need to be recalculated anytime other relationship edges were considered. To calculate GraSM shared information in real-time requires $O(k^3)$ operations. This result follows from the $O(k^2)$ calculations required to constructing the DCA and an added $O(k)$ cost for path counting at each step leading to a runtime that is cubic in the number of common ancestors.

Adjusted GraSM

Close study of the behavior of GraSM on real-world datasets leads to some unexpected results. In many cases, the DCA comprises a large proportion of the common ancestor set including the root node. An illustration of this behavior is provided in Supplemental Information Trace 1. GraSM is calculated by averaging the IC of DCA terms and may decrease as more shallow terms are included. In determining the membership of the DCA set, GraSM uses a *greater than or equal to*comparison on the number of paths. By changing the *greater than or equal to* (≥) to a *strictly greater than* (>) in the path number comparison, the algorithm's behavior changes and fewer terms are included in the DCA set. This modified algorithm is referred to as adjusted GraSM (A-GraSM). See Supplemental Information Trace 1 and 2 for further details on A-GraSM. The time complexity of the adjusted algorithm is the same as GraSM, $O(k^3)$.

Semantic frontier

Zhang and Lai introduced the exclusively inherited common ancestors set as an alternative to the DCA [21] that calculated a subset of the DCA in linear time. Our group developed an efficient implementation of the algorithm by Zhang and Lai, called the semantic frontier (SF) algorithm. A simple analogy can help to explain the exclusively inherited common ancestor set. The common ancestor set can be considered a region of the GO graph. Terms that form the *semantic frontier* lie on the frontier of the territory formed by the common ancestors. Specifically, terms in the semantic frontier set are common ancestor terms that have an incoming edge leading from one of the input terms not shared by paths leading from the other. The average IC of this set is returned as the shared information. The SF implementation offers a reduced search space over the algorithm proposed by Zhang and Lai. The SF algorithm is based on the concept of a breadth first search (BFS) visitor. A BFS visitor performs actions when certain events in a BFS occur. A BFS is performed for each of the two input terms and the SF set is calculated by examining the visited edges entering the common ancestor set. A detailed description of the SF algorithm is available in Appendix A.5.

The time complexity of the SF algorithm is $O(k)$. SF calculates the common ancestor set in $O(k)$, each BFS visitor operates in $O(k)$, and checking for frontier edges takes $O(k)$ time. Zhang and Lai report the complexity of their algorithm as $O(n\log n)$ [21]. The greater complexity reported by Zhang and Lai may be the result of less efficient ancestor or IC access which GGTK provides in $O(k)$ and $O(1)$ time respectively. Based on the increase in GO graph size in past 10 years, our early exit and reduced search space features may yield further performance gains in the future.

From shared information to term similarity

Methods for calculating term similarity can be constructed from shared information calculations through substitution into Resnik, Lin, and Jiang-Conrath (see Eqs. (6), (7), and (8)). In order to construct term similarity calculators using these method, GGTK provides a shared information interface that allows the modular combination of shared information and term similarity algorithms. This combination allows multiple distinct term similarity measures to be constructed. The interface promotes extensibility allowing other researchers to construct shared information algorithms and immediately combine them with existing term similarity measures.

Gene similarity from all-pairs of terms aggregation

The calculation of gene functional similarity relies on methods that aggregate the all-pairs term similarity between gene annotations [6]. GGTK provides *max*, *average*, and *best-match average* (BMA) gene similarity aggregators. Research by Pesquita et al. [19] and others has suggested that BMA outperforms other methods of aggregation. Based on initial experiments using the Collaborative Evaluation of Semantic Similarity Measures (CESSM) online tool, we arrived at the same conclusion. We evaluated the gene similarity benchmarks using BMA, but other aggregation methods may be useful in other contexts.

The number of gene-to-gene semantic similarity measures (N_{GSS}) explodes in a combinatorial fashion as described in Eq. (9) where $|si|$ is the number of shared information methods, $|ss|$ the number of semantic term similarity methods, $|a|$ the aggregation methods, and the final term, 3, represents the distinct ontologies, BP, MF, and CC. As new methods are devised to calculate term similarity, shared information, and aggregation of term similarity, the number of gene similarity measures grows combinatorially. This increasing makes evaluation a challenge, but offers a rich set of measures from which to choose. We explore whether these measures are significantly distinct in terms of performance and if gains can be achieved by using multiple measures.

$$N_{GSS}=|si|*|ss|*|a|*3 \qquad (9)$$

Jaccard-based gene similarity measures

Some gene similarity methods act on sets of terms without the need to explicitly calculate term-to-term similarity or employ aggregation of term similarity. To contrast the performance of shared information methods with other established approaches, we consider alternative gene similarity methods reported by the literature to perform well. Many of these approaches are variations of the Jaccard index [32] for set similarity. The Jaccard index calculates the ratio of the intersection to the union of sets. GGTK provides four Jaccard-based gene similarity measures. Gentleman [25]introduced a Jaccard-based gene similarity measure that calculates the ratio of shared ancestors between two terms to the union of each terms' ancestors. Let At be the set of all ancestors of term *t* (including *t*). The set At can be thought of as the induced subgraph of a term *t* in the ontology DAG. Eq. (10) describes the measure called SimUI, a similarity based on the union and intersection of term ancestors.

$$SimUI(t_1,t_2)=\frac{|A_{t_1}\cap A_{t_2}|}{|A_{t_1}\cup A_{t_2}|} \qquad (10)$$

Pesquita et al. [19] developed an information content weighted version of SimUI called SimGIC, a graph-based information content similarity. SimGIC is defined in Eq. (11).

$$SimGIC(t_1, t_2) = \frac{\sum_{t \in A_{t_1} \cap A_{t_2}} IC(t)}{\sum_{t \in A_{t_1} \cup A_{t_2}} IC(t)} \tag{11}$$

Studies by Mazandu et al. [26] introduced two modified methods similar to SimGIC. These methods are called by those authors, SimDIC and SimUIC after Dice and *universal* indexes. These measures are defined in Eqs. (12) and (13).

$$SimDIC(t_1, t_2) = \frac{2 * \sum_{t \in A_{t_1} \cap A_{t_2}} IC(t)}{\sum_{t \in A_{t_1}} IC(t) + \sum_{t \in A_{t_2}} IC(t)} \tag{12}$$

$$SimUIC(t_1, t_2) = \frac{\sum_{t \in A_{t_1} \cap A_{t_2}} IC(t)}{\max\left\{\sum_{t \in A_{t_1}} IC(t), \sum_{t \in A_{t_2}} IC(t)\right\}} \tag{13}$$

As these methods operate on the gene-level term sets, they avoid the all-pairs-of-terms calculation and need no aggregation step making these Jaccard-based methods more efficient than the shared information methods. The linear time shared information algorithms showed acceptable speed for all the practical applications of this work despite being necessarily slower than these Jaccard-based methods. For this reason, their execution time was not measured.

Decoupling term similarity calculation from gene similarity

As some shared information methods are computationally inefficient, we wanted to decouple the calculation of shared information from the calculation of gene similarity. To achieve efficient calculation of gene similarity, GGTK provides capabilities to generate and import pre-computed term similarity matrices. By pre-computing term similarity, term similarity algorithms can operate on sets of terms and access the similarity of a pair of terms in $O(1)$ time. Although the memory cost would appear to be $O(n^2)$ in the worst case, an optimization mitigates this cost by calculating only similarity between terms that appear in a given corpus (rather than all terms in the ontology). The separation of terms in each of the disjoint ontologies (BP, MF, CC) provides further space savings. If no gene in the corpus has a particular function annotation there is no need to calculate its similarity to other terms. These optimizations greatly reduce the memory cost and allow for efficient calculation of the gene-to-gene functional similarity.

Measuring execution speed of shared information algorithms

The calculation of term similarity matrices also serves as a benchmark for measuring each algorithms' execution speed. In order to quantify the execution speed of the algorithms in each complexity class, the wall-clock time of each matrix calculation was recorded for each shared information algorithm using the Linux *time* command. In calculating the matrix, only those terms with annotations in the human corpus need to be analyzed. Using this optimization, the term similarity matrices were calculated processing 11,394 annotated terms for BP, 4149 for MF, and 1546 terms for CC. Taking advantage of symmetry ($Sim(A,B) = Sim(B,A)$), the resulting number of calculations equals the number of elements in the upper triangular portion of the term similarity matrix $\left(\frac{n(n-1)}{2}\right)$. Where the all-pairs-of-terms similarity is calculated, 64,905,921 pairs were processed for the BP, 8,605,026 for MF, and 1,194,285 for CC. The resulting wall-clock time for each of these calculations is reported as the execution time for each method.

Semantic similarity performance evaluations

Various performance evaluations have been put forward starting with Lord et al. [11]who measured the Pearson correlation between gene-to-gene semantic similarity scores and sequence similarity. Foundational approaches accepted in the literature measure the correlation between gene-to-gene semantic similarity and sequence, domain set, or expression profile similarity [14]. Perhaps more useful evaluations measure the predictive power of semantic similarity to discover protein–protein interactions [6,29] or to correctly cluster genes belonging to known pathways [21]. Machine learning evaluations such as these have more applications in the field.

Few studies have addressed the inherent uncertainty in both GO graph structure and the limited depth and breadth of gene annotations [29]. The machine learning community has long used bootstrapping to improve generalization and to overcome issues of noise [33]. To address the issue of uncertainty, we employ robust bootstrapping approaches that provide generalized measures of performance as well as performance distributions that allow direct statistical comparisons between similarity methods. In the following sections we describe the specific performance evaluations and datasets used to compare a gene similarity method with respect to the shared information algorithms. In this work we evaluate five shared information algorithms using three term similarity measures across the three ontologies of GO resulting in 45 distinct gene similarity measures (5 shared information methods * 3 term similarity methods * 3 ontology types). In addition, these methods are compared to 12 gene similarity methods derived from the Jaccard-based methods (4 Jaccard-based methods * 3 ontology types). The performance of 57 measures in total has been analyzed.

CESSM

The Collaborative Evaluation of Semantic Similarity Measures (CESSM) [34] is an online dataset and comparison tool used to evaluate gene similarity measures. Though CESSM has some issues [27], the community has accepted it as a useful but limited performance benchmark which can provide comparisons with 11 other measures. The CESSM annotation and test data was downloaded and used to evaluate each measure. The 45 shared information methods were submitted individually to the CESSM server and the results, once collected, were analyzed. CESSM is available at xldb.di.fc.ul.pt/tools/cessm/. CESSM was used to evaluate term aggregation techniques. BMA performed best on this benchmark. The aggregators *max* and *average* performed poorly and were not considered for subsequent analysis. This agrees with similar observations by Pesquita et al. [19] and others. This analysis motivated the choice of BMA and the results for each ontology type are available in Supplemental Information.

Relative reciprocal BLAST score

The relative reciprocal BLAST score (RRBS) is a measure of sequence similarity developed by Pesquita et al. [19] derived from the BLAST alignment tool [35]. Eq. (14) gives the RRBS definition between two sequences A and B [19]. All-pairs BLAST was performed using an e-value cut off of 1e−4 after [19] with the human protein dataset. Pesquita et al. noted the relationship between shared information and RRBS is non-linear. For this reason, we evaluated the non-linear Spearman's rank correlation, or Spearman's ρ, between gene similarity values and the sequence alignment derived RRBS. The performance distribution of the Spearman correlation was calculated by taking 1000 bootstrap samples of size 100,000 from the set of all RRBS scores calculated.

$$RRBS(A,B) = \frac{BLAST_{bitscore}(A,B) + BLAST_{bitscore}(B,A)}{BLAST_{bitscore}(A,A) + BLAST_{bitscore}(B,B)} \qquad (14)$$

Jaccard index set similarity of Pfam domains

The Jaccard index [32] measures the similarity between two sets as the size of their intersection divided by the size of their union. In general, the functional domains, rather than just sequence, dictate a protein's function. To this end, Pfam domains [36]have been used in various gene similarity measure evaluations [13,20,27]. Taking genes as sets of domains (D_A and D_B), a domain similarity method is derived (Eq. (15)). On the human protein dataset, Pfam domains were predicted using HMMER3 [37] with an e-value threshold of $3e^{-6}$. After removing gene isoforms, the all-pairs Jaccard Pfam similarity was calculated for the protein dataset. As above, performance distributions for each algorithm were generated by calculating the Spearman correlation between the gene similarity measure and Jaccard Pfam for 1000 bootstrap samples of size 100,000.

$$Sim_{Jaccard}(D_A, D_B) = \frac{|D_A \cap D_B|}{|D_A \cup D_B|} \qquad (15)$$

TF–IDF cosine similarity of Pfam domains

Term frequency –inverse document frequency (TF–IDF) is a technique in information retrieval that weights terms by their relative specificity [38]. Similarly, domains that appear frequently in a diverse set of proteins may have little influence on the protein's function. Song et al. [39] previously applied TF–IDF to protein domains. Eq. (16) shows the weight for a particular domain d belonging to a protein, where f_d is the frequency of that domain in the protein, N is the total number of proteins in the corpus, and n_d is the number of proteins having the domain. From this weight measure, proteins are represented as a vectors of domain weights (A and B) and cosine similarity (Eq. (17)) represents the protein similarity. Performance distributions for this measure were calculated using Spearman correlation and bootstrapping as described above.

$$w_d = f_d * \log\left(\frac{N}{n_d}\right) \tag{16}$$

$$Sim_{TF-IDF}(A,B) = \frac{A \cdot B}{\| A \| \| B \|} \tag{17}$$

Gene expression across 79 human tissues

Highly correlated genes are often functionally related. Studies in the literature [4,23,28] have evaluated performance by measuring gene-to-gene semantic similarity correlation with gene expression correlation. The gene similarity measures were evaluated in terms of their correlation with microarray gene expression across 79 human tissues (NCBI GEO accession GSE1133). Probes from the human U133A array were mapped to their Refseq identifiers which were then mapped to Uniprot identifiers. Genes without any GO annotations in GOA were removed. After filtering, the all-pairs correlation of 5688 genes was calculated resulting in 16,173,828 unique correlation pairs. Both Pearson and Spearman gene correlations were calculated. As in previous literature [6], the absolute value of the correlation was calculated between expression pairs to attempt to detect a relationship either negative or positive. The semantic methods were then evaluated for their Spearman correlation to either the absolute Pearson or Spearman expression correlation. From these correlation pairs, 100 bootstrap samples were taken for each algorithm. Each sample had a size equal to 15% of the total number of pairs (>2.4 million).

Reactome pathway analysis

Uncovering pathway relationships between genes constitutes an important use case for semantic similarity algorithms that has been studied in the literature [3,5,21]. Using Reactome [40], we tested the performance of gene similarity methods in terms of their ability to partition sets of genes into known pathways by clustering. The variation of information (VI) criterion [41] was used to determine the agreement between known Reactome pathways and partitions of genes derived from hierarchical clustering using Ward's method [42] and gene-to-gene semantic similarity based distances (1 - similarity). Specifically, 100 datasets were generated from Reactome by randomly selecting 10 human pathways having between 10 and 150 proteins. These datasets were further processed replacing any overlapping pathways with non-overlapping pathways to remove any ambiguous assignments. For each gene similarity method, distance was calculated and the proteins were clustered into 10 groups. VI was calculated between the gene similarity based clustering and the known pathway assignments from Reactome. The 100 separate datasets were used to construct performance distributions.

Protein–protein interaction prediction

Several works in the literature [3,6,29] have used protein–protein interaction prediction to evaluate semantic similarity measures. Following previous methods, positive and negative datasets were generated from the Interologous Interaction Database (I2D) [43], and the semantic similarity measures were evaluated by their ability to distinguish interacting from non-interacting protein pairs. The I2D public version 2.3 was downloaded from http://ophid.utoronto.ca/ and used as the positive set of interacting proteins (228,847 human interactions). A negative dataset of equal size was generated after Guo et al. [3] by randomly choosing protein pairs resulting in a balanced dataset of 457,694 interactions. For each gene similarity method, 100 bootstrap samples of a size equal to 15% of the original dataset were used to calculate receiver operator characteristic (ROC) curves using the ROCR package [44]of the R programming language. A ROC curve plots the true positive rate (TPR), the ratio of true predictions to all predictions made, against the false positive rate (FPR), the ratio of false predictions to all prediction, across a range of different thresholds. For the gene-to-gene semantic similarity measures, the ROC curve would be equivalent to sorting all the values of semantic similarity and counting the number of true and false predictions below each unique value for plotting. From these 100 samples, a performance distribution of the area under the ROC curve (AUC) values was used to statistically compare the performance of each semantic similarity measure. A classifier must have an AUC above 50% to be considered better than random guessing.

Isolating the effect of shared information from other factors

The choice of shared information, term similarity algorithm, and ontology type specify the gene-to-gene semantic similarity measure. To determine the effects of shared information on performance in the previous evaluations it is important to isolate the effects of other factors that influence the results.

Regression analysis of mean performance

Linear regression analysis on the bootstrapped mean performance values is used to examine the effects of shared information type, term similarity algorithm, and gene ontology type. A design matrix was created with each row representing a separate gene-to-gene semantic similarity method constructed through combining the different factors under study. Each column of the design matrix represents the different factors as categorical variables, with the ontology variable taking a value in {*BP*, *MF*, *CC*}, the term similarity variable taking a value in {*Resnik*, *Lin*, *JC*}, and the shared information variable taking a value in {*casi*, *mica*, *grasm*, *agrasm*, *sf*}. Using this design matrix, linear models were trained for each performance evaluation using the mean bootstrapped performance as the response variable. The models where creating using the *lm* function of the R programming language. The influence of each factor is reported as the negative log of the regression p-value taken from the analysis of variance of each fitted model using the *anova* function of the R programming language. Separate models were fitted for each of the six performance benchmarks and the relative influence of each factor is reported for all evaluations.

Performance ranking and statistical ties

Within each choice of term similarity and ontology type, the shared information algorithms were ranked based on statistical tests of their performance. For a specific choice of term similarity and ontology type, the shared information methods were sorted based on their mean performance and ranked using a statistical method operating on their performance distributions. A statistical tie in performance between two methods is determined if a t-test of the methods' performance distributions fails to reject the null hypothesis of equal means (p-value > 0.05, Welch's two sample t-test). The ranks are exhausted meaning that if two methods are tied for first place, method 1 and 2 receive the rank of 1, but the next best performer receives the rank of 3 (rather than 2). The overall ranks of all shared information methods are reported for every combination of term similarity and ontology type. The average shared information algorithm rank is reported for each ontology type and term similarity method.

Ensemble classifiers and cross-validation

With such a large number of gene similarity approaches, could they be combined through ensemble methods to improve performance on machine learning tasks? To answer this question, we applied majority voting to the task of protein-protein interaction prediction. A majority voting classifier makes a prediction based on the votes of a panel of other classifiers [45]. Although the area under the ROC curve provides a good estimate of a classifier's behavior across a range of thresholds, in practice, a single threshold must be chosen for a particular task. A process that maximized the F-score over a set of training data selected the threshold for an individual gene-to-gene semantic similarity method to be used as a classifier. The F-score represents the trade off between making true predictions and limiting false predictions. It is the harmonic mean of precision and recall and is defined in Eqs. (18), (19), and (20). These values were calculated using the ROCR [44] package in R. This training process was repeated for every individual method. From these individual classifiers, voting classifiers were constructed that predicted an interaction only if a majority of its constituent classifiers also predicted an interaction. Three ontology specific voting classifier were designed to include only methods in BP, MF, or CC categories. These classifiers helped to determine if any advantage could be derived from ensemble classifiers limited to a single GO type. A final voting predictor was constructed which consisted of all individual methods.

$$\text{precision} = \frac{\text{true positives}}{\text{true positives} + \text{false positives}} \tag{18}$$

$$\text{recall} = \frac{\text{true positives}}{\text{true positives} + \text{false negatives}} \tag{19}$$

$$\text{F-score} = 2 * \frac{\text{precision} * \text{recall}}{\text{precision} + \text{recall}} \tag{20}$$

Ten-fold cross-validation was employed to test the performance of all individual classifiers as well as the four ensemble classifiers. Using the protein interaction dataset from Section 2.5.7, ten random partitions, or folds, were constructed each containing approximately 10% of the original 457,694 interactions. The classifiers were trained using nine folds and tested for the percent of misclassified instances (error rate) on the remaining fold. This process was repeated ten times. The classification error is reported for all 57 methods including the four voting classifiers.

RESULTS

Execution time performance of the shared information algorithms

The all-pairs term similarity matrix for annotated terms in BP, MF, and CC was calculated for each shared information method and the execution times of these calculations were recorded. The wall-clock execution times for these calculations showed, as expected, that GraSM and A-GraSM methods are orders of magnitude slower than the other shared information methods. Fig. 2 shows the execution times of all shared information methods combined with Resnik term similarity. The other term similarity measures showed comparable execution times. The complete list of execution times is available. Both GraSM and A-GraSM have $O(k^3)$ time complexity, and their execution times dwarf the runtimes of the other algorithms. The execution time of the SF algorithm is greater than CASI or MICA. Both CASI and MICA proved to be the most efficient algorithms in terms of execution speed. The term similarity matrices for BP, MF, and CC required 64.91 million, 8.60 million, and 1.19 million term pair calculations respectively. Interestingly, despite requiring over 8 million more calculations, the MF processes completed faster than the CC processes. These results suggest that the topology of the GO graph plays an important role in determining the execution speed of semantic algorithms and that functions of the raw number of terms in an ontology may not accurately reflect their complexity. Fig. 1supports this conclusion since the branching factor and average number of ancestors of the MF ontology is less than that of the CC ontology despite MF having more terms. The Jaccard-based methods do not require calculating term-to-term similarity so this evaluation does not apply. The Jaccard-based term-set level measures are known to be more efficient. These findings illustrate that the increased time complexity of the GraSM methods could be computationally prohibitive in some situations, and the problem may worsen as GO graph complexity grows.

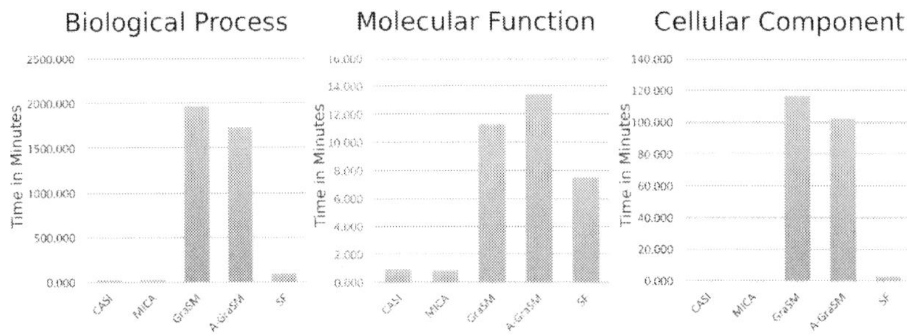

Figure 2: The execution times for the all-pairs term similarity for Resnik term similarity show that GraSM and A-GraSM methods are slower than other shared information methods.

RRBS: sequence similarity

Using the 519,892 protein pairs that passed selection, Spearman's ρ correlation between RRBS and each of the 45 shared information-based and 12 Jaccard-based gene similarity measures was evaluated by the bootstrapping method described in the Section 2.5.2. The correlation distributions of all measures are shown in Fig. 3. Supplemental Information provides the mean performance and standard deviations for all measures. The measures are organized by term similarity type and ontology type. In general, BP methods perform best in terms of RRBS correlation. For Lin and Jiang-Conrath methods no dramatic differences are observed between shared information algorithms. For the Resnik methods, CASI and MICA lag behind GraSM, A-GraSM, and SF across all ontology types. With the CC ontology, the CASI and MICA Resnik methods show dramatically lower correlation than all other methods. Despite poor and average performance in the CC and MF ontologies respectively, Resnik BP methods using GraSM, A-GraSM, and SF show better correlation with

RRBS than any other methods. Of the MF methods, Jiang-Conrath term similarity shows the lowest RRBS correlation among the term similarity algorithms. The Jaccard-based methods perform best out of the MF methods. The correlation of the MF Jaccard methods roughly equals the performance of the BP Jaccard methods. In the BP ontology, the Jaccard methods showed similar performance to the Jiang-Conrath methods but were lower than Lin and the best Resnik methods. These results show that BP ontology methods most closely correlate with RRBS similarity. Consistent with the results of Pesquita et al. [19], SimGIC performed well among the MF methods. In that work the authors did not consider BP or CC based methods citing the works Lord et al. [11,12] that report a loose correlation between gene semantic similarity and sequence similarity. Jain and Bader [4] analyzed the correlation between semantic methods and sequence similarity in all ontologies, and their results show the CC methods have a higher sequence correlation than MF or BP. From the results of Mazandu et al. [26], BP and CC methods appear to show better correlation with sequence than MF. As Pesquita et al. [19] demonstrated the relationship between semantic similarity and sequence similarity is not linear, making the Spearman correlation a more appropriate measure. The use of Spearman correlation over Pearson and using more recent annotation and GO graph data likely account for some of the differences with previous works.

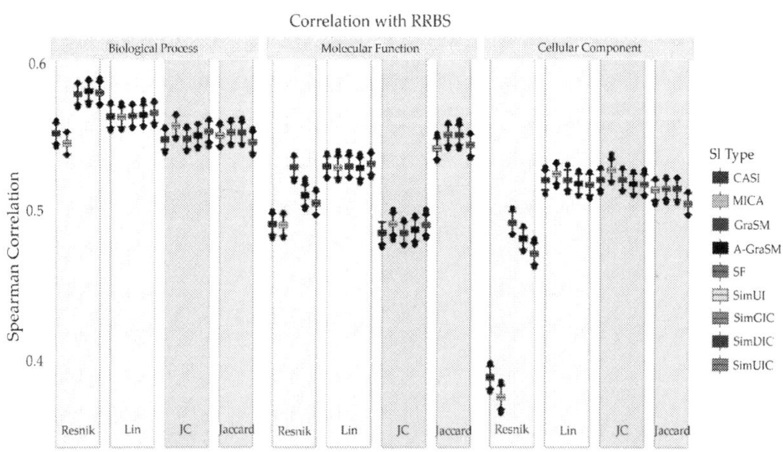

Figure 3: Performance distributions for the BLAST-based RRBS benchmark for all 57 measures organized by term similarity algorithm and ontology type.

Table 1: Ranking of shared information for the BLAST-based RRBS benchmark against ontology type and term similarity type. Methods with greater correlation have lower rank. Bold font represents the best average rank for each category.

	Biological process				**Molecular function**				**Cellular component**				**Mean by term similarity**		
Method	**Resnik**	**Lin**	**JC**	**Mean**	**Resnik**	**Lin**	**JC**	**Mean**	**Resnik**	**Lin**	**JC**	**Mean**	**Resnik**	**Lin**	**JC**
CASI	4	3	5	4.00	4	2	4	3.33	4	2	2	2.67	4.00	**2.33**	3.67
MICA	5	5	1	3.67	5	4	1	3.33	5	1	1	2.33	5.00	3.33	**1.00**
GraSM	3	3	4	3.33	1	2	4	2.33	1	2	2	**1.67**	**1.67**	**2.33**	3.33
A-GraSM	1	2	3	2.00	2	5	3	3.33	2	4	4	3.33	**1.67**	3.67	3.33
SF	2	1	2	**1.67**	3	1	2	**2.00**	3	5	5	4.33	2.67	**2.33**	3.00

Statistical comparisons between performance distributions examine the relative performance between the methods and establish a ranking. This approach to ranking is described in Section 2.6.2. Table 1 shows the rankings for all shared information methods. These data give insight into which shared information algorithms show statically significant differences in performance for each combination of term similarity and ontology type. SF shared information works best with BP and MF methods. CC methods work well with GraSM on the RRBS benchmark. GraSM and A-GraSM show good performance with Resnik. Lin based method showed CASI, GraSM, and SF as the best shared information methods. MICA shared information ranked first in all JC methods.

Jaccard index of Pfam domain sets

The all-pairs Pfam domain similarity was calculated with the Jaccard index resulting in 1,219,558 protein pairs. The performance distributions are shown in Fig. 4 depicting all 57 measures organized by term similarity type and ontology type. The complete listing of mean correlation performance and standard deviations can be found in Supplemental Information. As with the RRBS dataset, the shared information methods show smaller variability in performance compared to differences in ontology type and term similarity algorithm. The CASI and MICA methods combined with Resnik term similarity again perform poorly compared to the other methods, but these shared information method are comparable with GraSM, A-GraSM, and SF when combined with Lin and Jiang-Conrath. The CC ontology methods lag the performance of MF and BP in terms of correlation with Pfam Jaccard similarity. The Jaccard-based gene similarity methods perform best with MF annotations, with the best method being SimUI. These methods perform slightly better in MF than in BP. Overall MF Lin methods performed best, but BP Resnik methods combined with GraSM, A-GraSM, and SF outperformed most other methods. As the worst performing ontology group, the CC ontology may fail to encode detailed domain or structural information. Despite having a higher correlation than the CC ontology methods, the BP and MF correlations were lower than in the RRBS benchmark with the best methods only achieving a correlation near 0.35.

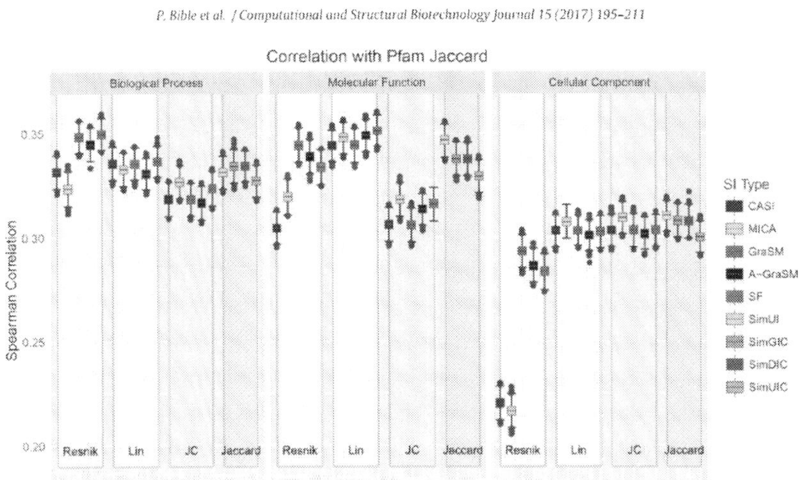

Figure 4: Performance distributions for the Pfam Jaccard benchmark for all 57 measures organized by term similarity algorithm and ontology type.

Table 2 shows the ranking of shared information methods for the Pfam Jaccard evaluations. Among the ontologies, SF performs best with BP and MF as in the RRBS evaluations. GraSM again shows the best performance for CC methods. Resnik, Lin and Jiang-Conrath method perform best with GraSM, SF, and MCIA respectively on the Pfam Jaccard dataset.

Table 2: Pfam Jaccard ranking of shared information against ontology type and term similarity type. Methods with greater correlation have lower rank. Bold font represents the best average rank for each category.

	Biological process				Molecular function				Cellular component				Mean by term similarity		
Method	**Resnik**	**Lin**	**JC**	**Mean**	**Resnik**	**Lin**	**JC**	**Mean**	**Resnik**	**Lin**	**JC**	**Mean**	**Resnik**	**Lin**	**JC**
CASI	4	2	3	3.00	5	4	4	4.33	4	2	4	3.33	4.33	2.67	3.67
MICA	5	4	1	3.33	4	3	1	2.67	5	1	1	2.33	4.67	2.67	**1.00**
GraSM	2	2	3	2.33	1	4	5	3.33	1	2	2	**1.67**	**1.33**	2.67	3.33
A-GraSM	3	5	5	4.33	2	2	3	2.33	2	5	5	4.00	2.33	4.00	4.33
SF	1	1	2	**1.33**	3	1	2	**2.00**	3	4	2	3.00	2.33	**2.00**	2.00

TF–IDF cosine similarity of Pfam domain sets

To correct for uninformative domains, we applied TF–IDF to proteins represented as sets of Pfam domains. With 1,219,558 unique protein pairs, evaluations were conducted using the bootstrapping method. Fig. 5 shows the distribution of Spearman correlations against TF–IF domain similarity (mean and standard deviation data are available in Supplemental Information). Again, shared information methods show less variability with the exception of Resnik methods using CASI and MICA. CASI and MICA Resnik methods are dramatically low in the CC ontology and show the lowest correlations of all measures. BP Resnik methods using GraSM, A-GraSM, and SF outperform all other methods in terms of TF–IDF correlation. The Jaccard-based methods perform similarly in BP and CC ontology, but these methods are among the worst performing measures in MF. The CC Jaccard-based methods show greater correlation with TF–IDF than the other CC methods. The BP methods preform best on average in this evaluations. These novels findings suggest that TF–IDF can be successfully applied to protein domain sets to correct for nonspecific domains and potentially uncover relationships in both biological processes and molecular functions.

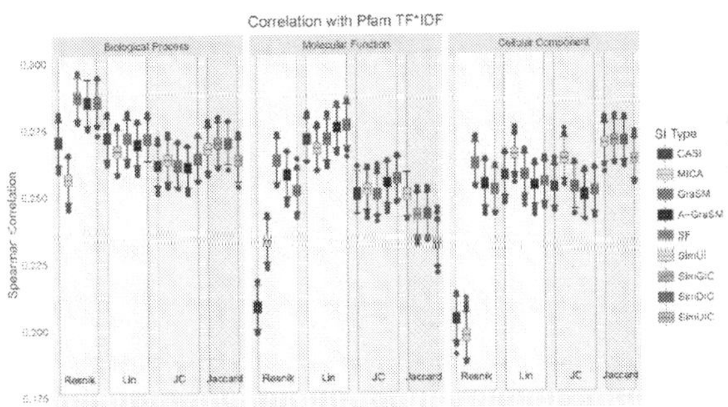

Figure 5: Performance distributions for the Pfam TF–IDF benchmark for all 57 measures organized by term similarity algorithm and ontology type.

The performance rankings of the shared information methods in the TF–IDF benchmark are shown in Table 3. With BP and CC methods in the TF–IDF benchmark, GraSM shows the best average ranking. SF shared information works best with MF methods in this dataset. Resnik methods work best with GraSM which is ranked first in all Resnik evaluations. Lin methods show the best correlations using CASI and GraSM while Jiang-Conrath methods perform best with MICA and SF shared information.

Table 3: Pfam TF–IDF ranking of shared information by term similarity type and ontology. Methods with greater correlation have lower rank. Bold font represents the best average rank for each category.

	Biological process				**Molecular function**				**Cellular component**				**Mean by term similarity**		
Method	**Resnik**	**Lin**	**JC**	**Mean**	**Resnik**	**Lin**	**JC**	**Mean**	**Resnik**	**Lin**	**JC**	**Mean**	**Resnik**	**Lin**	**JC**
CASI	4	1	3	2.67	5	3	4	4.00	4	2	2	2.67	4.33	**2.00**	3.00
MICA	5	5	2	4.00	4	5	3	4.00	5	1	1	2.33	4.67	3.67	**2.00**
GraSM	1	1	3	**1.67**	1	3	4	2.67	1	2	2	**1.67**	**1.00**	**2.00**	3.00
A-GraSM	2	4	5	3.67	2	2	2	2.00	2	5	5	4.00	2.00	3.67	4.00
SF	2	3	1	2.00	3	1	1	**1.67**	3	4	4	3.67	2.67	2.67	**2.00**

Gene expression across 79 tissues

All 57 measures were evaluated based on their correlation with gene expression correlation. Fig. 6 shows the absolute gene expression correlation for all methods under study. The complete mean and standard deviation data are available in Supplementary Information (Pearson gene correlation) and (Spearman). BP Jiang-Conrath methods show the best performance at slightly greater than 0. All other methods show a negative correlation with expression correlation. These findings are consistent with the work of Xu et al. [6] which found that global gene expression correlates poorly with semantic similarity. Xu et al. observed a steady increase in semantic similarity as gene expression pairs are binned into sets of highly correlated gene sets. The same trend is confirmed in our data. At such low levels of correlation, one might expect greater variability in performance due to a lack of any strong relationship. The data in Fig. 6 surprisingly shows tighter distributions than in other evaluations. The differences in performance between shared information types are minimal, with the exception of MICA.

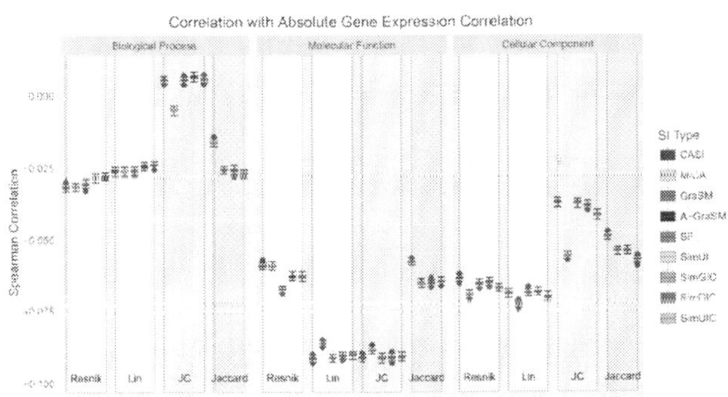

Figure 6: Performance distributions for absolute gene expression correlation (Pearson) against all 57 measures organized by term similarity algorithm and ontology type.

The shared information rankings are provided in Table 4 for absolute gene expression. BP, MF, and CC methods work best with SF, MICA, and CASI shared information respectively for gene expression. CASI performs best with Resnik methods and A-GraSM with Lin methods. CASI, GraSM, and A-Grasm work well with Jiang-Conrath methods. The low and negative correlations associated with this benchmark makes the significance of this ranking less clear.

Table 4: Absolute gene correlation benchmark ranking for shared information methods organized by term similarity type and ontology. Methods with greater correlation have lower rank. Bold font represents the best average rank for each category.

	Biological process				**Molecular function**				**Cellular component**				**Mean by term similarity**		
Method	**Resnik**	**Lin**	**JC**	**Mean**	**Resnik**	**Lin**	**JC**	**Mean**	**Resnik**	**Lin**	**JC**	**Mean**	**Resnik**	**Lin**	**JC**
CASI	4	3	3	3.33	1	5	3	3.00	1	2	1	**1.33**	**2.00**	3.33	**2.33**
MICA	4	3	5	4.00	1	1	1	**1.00**	5	5	5	5.00	3.33	3.00	3.67
GraSM	3	3	3	3.00	5	4	3	4.00	3	2	1	2.00	3.67	3.00	**2.33**
A-GraSM	2	2	1	1.67	3	3	3	3.00	2	1	3	2.00	2.33	**2.00**	**2.33**
SF	1	1	2	**1.33**	3	2	2	2.33	4	4	4	4.00	2.67	2.33	2.67

Reactome clustering

Using 100 randomly generated pathway datasets (see Section 2.5.6), the mean VI similarity (1 - normalized VI distance) was used to compare the closeness of semantic gene similarity derived clusters to their known Reactome assignments. Fig. 7 shows the VI similarity distribution for all 57 methods tested. The performance variability on this benchmark is much higher than in other evaluations. The BP methods outperform those of MF and CC, but within each ontology type all shared information based methods are essentially equivalent regardless of term similarity type. The Jaccard-based method tend to underperform the other methods. As the Reactome pathways were chosen randomly, the high variability in performance is expected. The success of BP methods confirms expectations since the BP ontology captures functional information most closely associated with biological pathways. The CC based methods out-perform MF method on average for this benchmark. The complete results are available in Supplemental Information.

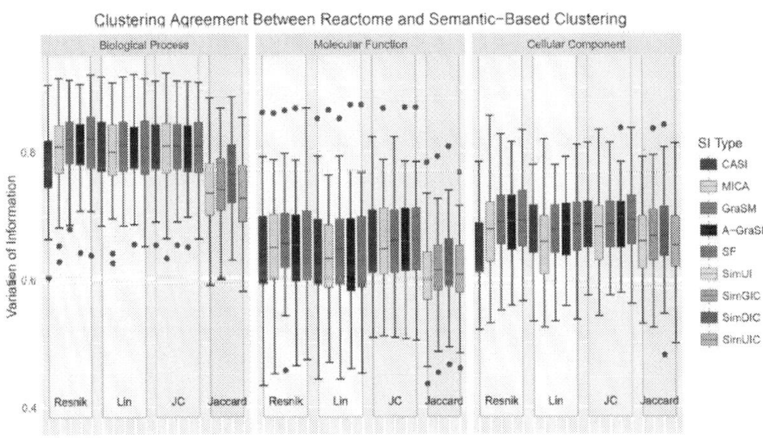

Figure 7: Performance distributions for Reactome clustering compared to the 57 gene similarity semantic measures organized by term similarity algorithm and ontology type.

Table 5: Reactome based statistical ranking for shared information methods organized by term similarity type and ontology. Methods showing a higher VI similarity to Reactome pathways have lower rank. Bold font represents the best average rank for each category.

	Biological process				**Molecular function**				**Cellular component**				**Mean by term similarity**		
Method	**Resnik**	**Lin**	**JC**	**Mean**	**Resnik**	**Lin**	**JC**	**Mean**	**Resnik**	**Lin**	**JC**	**Mean**	**Resnik**	**Lin**	**JC**
CASI	5	1	1	2.33	1	1	1	**1.00**	5	1	1	2.33	3.67	**1.00**	**1.00**
MICA	1	1	1	**1.00**	1	1	1	**1.00**	4	5	5	4.67	2.00	2.33	2.33
GraSM	1	1	1	**1.00**	1	1	1	**1.00**	1	1	1	**1.00**	**1.00**	**1.00**	**1.00**
A-GraSM	1	1	1	**1.00**	1	1	1	**1.00**	1	1	1	**1.00**	**1.00**	**1.00**	**1.00**
SF	1	1	1	**1.00**	1	1	1	**1.00**	1	1	1	**1.00**	**1.00**	**1.00**	**1.00**

Table 5 gives the performance rankings of the shared information methods for the Reactome benchmark. As the Reactome results show high variability, most methods could not be determined to be statistically different from one another. All shared information methods within each term similarity group proved to be equivalent in the MF ontology in terms of the VI similarity to Reactome pathway clusters. In BP, the CASI method performs poorly with Resnik. In the CC ontology, CASI and MICA perform poorly with MICA particular ill suited to the Lin and Jiang-Conrath methods. Resnik methods show poor performance with CASI and MICA, while MICA Lin and Jiang-Conrath methods show poor performance reflecting issues in combination with the CC ontology.

Protein–protein interaction prediction

The predictive power of each gene similarity measure was evaluated on the human I2D interaction dataset. Performance distributions of the AUC (described in Section 2.5.7) appear in Fig. 8. Complete data for the mean and standard deviation of the AUC are available in Supplemental Information. This prediction-based evaluation showed heightened variability between shared information methods compared to other datasets, especially with the Jiang-Conrath methods. Within Jiang-Conrath term similarity, MICA shared information performed best.

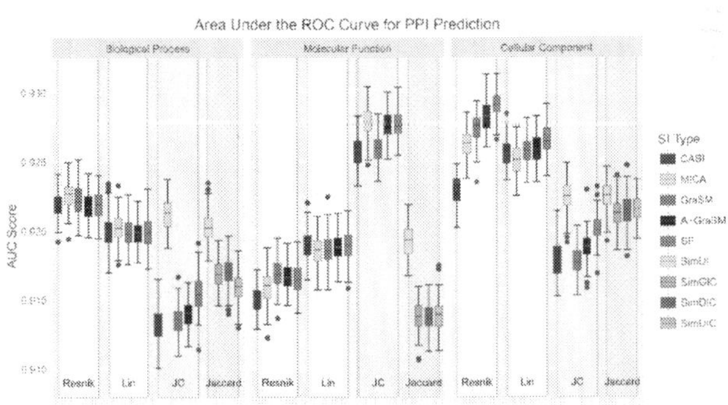

Figure 8: Performance distributions for protein–protein interaction prediction by area under the ROC curve for the 57 gene similarity semantic measures organized by term similarity algorithm and ontology type.

Table 6: Protein interaction prediction ranking for shared information methods organized by term similarity type and ontology. Rankings were established by comparisons based on the area under the ROC curve calculated from bootstrap samples. Methods with greater AUC scores have lower rank. Bold font represents the best average rank for each category.

	Biological process				**Molecular function**				**Cellular component**				**Mean by term similarity**		
Method	**Resnik**	**Lin**	**JC**	**Mean**	**Resnik**	**Lin**	**JC**	**Mean**	**Resnik**	**Lin**	**JC**	**Mean**	**Resnik**	**Lin**	**JC**
CASI	3	1	4	2.67	5	1	4	3.33	5	2	4	3.67	4.33	**1.33**	4.00
MICA	1	1	1	**1.00**	4	4	1	3.00	4	5	1	3.33	3.00	3.33	**1.00**
GraSM	2	3	4	3.00	1	4	4	3.00	3	2	4	3.00	**2.00**	3.00	4.00
A-GraSM	3	3	3	3.00	1	1	1	**1.00**	2	2	3	2.33	**2.00**	2.00	2.33
SF	3	3	2	2.67	3	1	1	1.67	1	1	2	**1.33**	2.33	1.67	1.67

The shared information methods varied greatly within CC Reskin with the SF and A-GraSM versions giving the best AUC values of all methods. The Jaccard-based methods show the worst performance among MF methods and are superior to only the Jiang-Conrath methods in the BP and CC ontologies. BP and MF methods tend to under perform compared to CC methods with the exception of MF Jiang-Conrath. This result is at odds with older evaluations such as Guo et al. [3] that found BP methods are better predictors than CC methods. These differences are likely due to the smaller dataset used in their analysis as well as the changes in GO structure and annotations in recent years. Conflicting performance reports in the literature illustrate the need for transparent open source tools that can evaluate these methods on a level playing field using consistent ontologies and annotations.

The performance rankings for the protein–protein interaction prediction benchmark is shown in Table 6. Based on the distribution of the area under the ROC curve values, the shared information methods were statistically compared. MICA, A-GraSM, and SF shared information perform best for BP, MF, and CC ontologies respectively. GraSM and A-GraSM perform well for Resnik on this benchmark. CASI is the best performer for Lin term similarity, and MICA works best for JC across all the ontologies in these evaluations.

Analysis of factors affecting semantic similarity performance

Using the six performance benchmarks described in the previous section, regression models were trained to assess the influence of ontology type, term similarity type, and shared information type on the mean performance of the shared information methods. A design matrix of categorical variables was constructed and used to create a linear model of mean performance (described in Section 2.6.1). Fig. 9 presents the influence of each factor on each benchmark as the negative log of the regression p-value. This data demonstrates that the choice of ontology contributes most to the variability in performance followed by the choice of term similarity, and shared information contributes the least. All evaluations tested uphold this trend. The Reactome benchmark shows the strongest influence by ontology followed by the gene expression evaluations. Ontology type exerts the smallest influence on TF–IDF and PPI benchmarks; however, this effect is still much greater than the other two factors. The effects of term similarity and shared information choice are small across all evaluations except PPI where the effects are negligible or zero. These results clearly show that the choice of ontology greatly effects the performance of gene similarity methods.

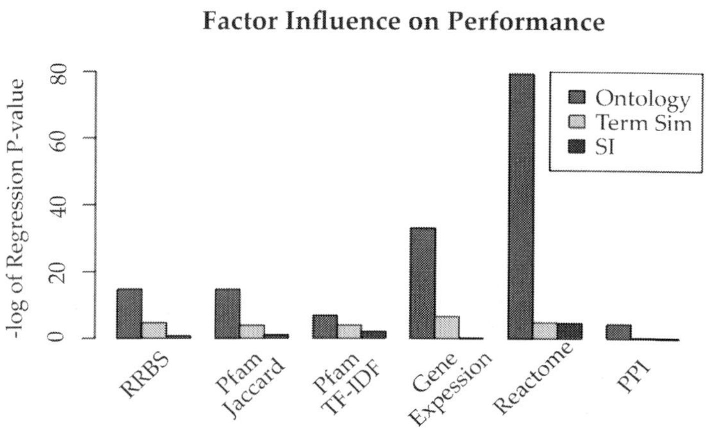

Figure 9: The relative influence of ontology type, term similarity method, and shared information type (SI) on the mean performance across six evaluations.

Leveraging semantic methods with majority voting

Using the protein–protein prediction dataset from the previous evaluations, we addressed the feasibility of combining semantic similarity methods to improve prediction. Using the 10-fold cross-validation method (described in Section 2.7), we analyzed the performance of all 57 measures in terms of their mean prediction error. For this evaluation, the percent of misclassified instances represents the classification error of each gene similarity semantic measure and voting classifier. Fig. 10 presents the cross-validation classification error for all methods organized by ontology type and term similarity method (lower is better). The ontology specific voting classifier performed equivalently with other methods within the ontology (the far right classifier within each ontology group in Fig. 10). Other single classifiers outperform the ontology specific voting predictor within each ontology. This result indicates that simple majority voting offers no advantage over the best methods within an ontology group; however, the combination of all semantic similarity methods substantially out performs even the best individual classifier. The failure of ontology specific voting predictors to confer any advantage for prediction suggests that variation among shared information or term similarity semantic methods alone cannot improve learning at least for this task. Analysis of the factors affecting semantic similarity performance of the previous section support this conclusion. The cross-validation evaluation show that BP methods tend to perform worse than all MF methods and worse than all but Jiang-Conrath CC methods. Interestingly, the MF methods perform best in cross-validation analysis while the CC methods tend to perform best in terms of the area under the ROC curve calculated through bootstrapping. These results indicate that the AUC metric may not totally capture the usefulness of a measure for classification.

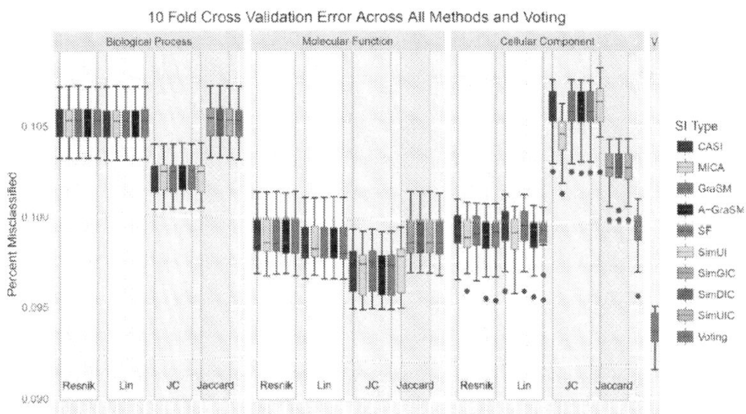

Figure 10: The percent of misclassified samples for each method under study, trained as classifiers, and four voting predictors evaluated by 10-fold cross-validation. The voting predictor for an ontology type is presented as the last classifier within that ontology (red), and the voting predictor utilizing all semantic methods is presented at the far right.

DISCUSSION

As semantic similarity measures have grown in popularity, a large number of measures have been developed making exhaustive performance assessment a growing challenge. IC based methods use measures of shared information to calculate term similarity and ultimately gene similarity. Recent works have devised alternative methods for calculating shared information. Through modular combination of shared information measures, term similarity calculations, and ontology choice, the number of semantic gene similarity measures explodes creating obvious challenges for performance evaluations. In a departure from previous works, we put forward robust methods to statistically compare semantic gene similarity measures in a manner that captures their generalized performance. We apply these methods to conduct a thorough investigation into the behavior of varying shared information measures. Given the large number of measure that can be created though modifications to the shared information, we considered the feasibility of leveraging these measures in ensemble classifiers for prediction gains. By isolating the effects of ontology type, we determined that the shared information algorithms themselves could not be combined to improve protein-protein interaction prediction. Methods across the distinct ontologies can be leveraged to improve prediction.

Xu et al. [6] combined information from across ontologies to improve prediction; however, they incorporated an artificial root node connecting BP, MF, and CC into a single graph. Their research found that BP methods offered the best performance on a yeast dataset. Later Jain and Bader [4] also showed BP to out perform MF and CC methods in interaction prediction in yeast based on area under the ROC curve analysis both with and without the inclusion of electronically inferred annotations. While our work focused on human protein–protein interactions, the results resemble the more recent work of Yang et al. [29] in yeast where they found CC to perform better than other methods. Based on ROC analysis, our results showed CC methods performing best. The work by Yang et al. cites Collins et al. [46] in identifying that much of the ROC curve represents classification thresholds that would be useless in practice for protein interaction prediction. This results from thresholds that admit far too many false positive predictions to be useful. This issue became evident in our analysis when many of the MF predictors showed strong performance based on cross-validation evaluation but less competitive performance when only the area under the ROC curve was considered.

The work by Mazandu et al. [26] more closely relates to this study as they used human interactions to assess the predictive power of semantic algorithms. The data set used by Mazandu et al. was comprised of a much smaller set of roughly 5000 curated interaction which were filtered to all contain BP and CC annotations. Citing their previous work [47], Mazandu et al. excluded MF based methods in their evaluation, and other work by Mazandu et al. [48] supported the exclusion of Jiang-Conrath methods. These caveats make a direct comparison to their work difficult. Mazandu et al. examined a large number of diverse measures which are out of score for this study of shared information methods. The shared methods among the two studies are the best match average (BMA) versions of Resnik (called RBMA, by Mazandu et al.), Lin (LBMA), SimUI, SimGIC (AGIC), SimDIC (ADIC), and SimUIC (AUIC). The area under the ROC curve reported Mazandu et al. for the CC base methods is 0.9999656 (RBMA), 0.4853167 (LBMA), 0.8483416 (SimUI), 0.9173889 (AGIC), 0.8486233 (ADIC), and 0.9654985 (AUIC). The values reported for the BP verions of these methods are 0.9995277 (RBMA), 0.6194642 (LBMA), 0.9582268 (SimUI), 0.9689432 (AGIC), 0.9514534 (ADIC), and 0.9654985 (AUIC). The range of these values is consistent with those found in this study for the area under the ROC curve benchmark. All the methods tested achieved high performance using the most up-to-date annotations and GO graphs. Inconsistent with the reports of this study is the disparity between Resnik and Lin methods within the same ontology. Based on our evaluations, we would expect these methods to have scored more closely in the evaluations of Mazandu et al.

Difficulty in comparing the results of semantic similarity analysis in GO is a known problem [14]. In an excellent review, Mazandu et al. [15] describe two key challenges known as the *dataset issue*, where different tools use different version of GO or annotation datasets, and the *scaling issue* that results from tools making different assumption regarding normalization methods and other minor considerations such as root membership in ancestor sets etc. GGTK is an attempt to provide correct, transparent, and modular implementations of semantic algorithms where the dataset issue can be tackled easily and the assumptions that could lead to the scaling issue are clearly stated. The benchmarks used in this work are provided as simple lists of protein pairs and scores to facilitate comparisons by other researchers (available at github.com/paulbible/ggtk). This work demonstrates that many conclusions from the literature still hold despite changes in the structure of the ontologies and increases in the number of annotations. In opposition to previous reports, we find that MF methods performed best in predicting protein interactions using cross-validation. We found that an ensemble of gene similarity semantic methods could out perform any single method in interaction prediction, but this boost in performance results mainly from the integration of distinct information gained through the separate ontologies of BP, MF, and CC. Furthermore, we have developed an efficient C++ toolkit with an easy-to-use Python interface, GGTK, that will not only allow other researchers to take advantage of these findings but also facilitate their own research into semantic methods.

ACKNOWLEDGMENTS

This work is supported by the National Basic Research Program of China2015CB964601 and 2013CB967001, the Li Foundation Heritage Prize, the 1000 Young Talent Plan China, the National Natural Science Foundation of China81570828, and the NSFC-Guangdong Fund for Application of Supercomputing using the Tianhe 2 Supercomputer to L. Wei. This study also utilized the high-performance computational capabilities of the Helix Systems at the National Institutes of Health, Bethesda, MD (http://helix.nih.gov). Additional support provided by the Intramural Research Program and the Office of Science and Technology of the National Institute of Arthritis and Musculoskeletal and Skin Diseases of the National Institutes of Health.

APPENDIX A. DETAILED DESCRIPTION OF SHARED INFORMATION ALGORITHMS

A.1. Common ancestor shared information

Common ancestor shared information (CASI) is a naive baseline algorithms that calculated the simple mean IC of all common ancestors. Eq. (A.2) defines CASI where $|CA(t_1,t_2)|$ is the number of shared ancestors.

$$SI_{CASI}(t_1,t_2) = \frac{\sum_{t \in CA(t_1,t_2)} IC(t)}{|CA(t_1,t_2)|} \quad \text{(A.1)}$$

A.2. Most informative common ancestor shared information

Most informative common ancestor (MICA) shared information is equivalent to Resnik term similarity and is defined by the IC of minimum subsumer. MICA shared information is completely described in the main text above through Eqs. (3), (4), and (5). The MICA represents the most specific concept that contains both t_1 and t_2 as descendants.

A.3. Couto et al. 2007, GraSM

The graph-based semantic measure or GraSM was developed by Couto et al. [20]. GraSM calculates the set of disjunctive common ancestors (DCA) and computes their average as the shared information. Eq. (A.2) defines the disjunctive ancestor pairs for a term *t* where a_1 and a_2 are ancestor terms and *Paths* represents the number of unique paths from the ancestor to the descendant [49]. Couto et al. [20] give an algorithm for calculating if a pair of ancestor terms (a_1,a_2) are disjunctive in a descendant term *t* reproduced here in Algorithm 1. From this definition, the DCA set is defined in Eq. (A.3). Finally, the shared information is calculated as the mean IC of the set of $DCA(t_1,t_2)$ in Eq. (A.4).

$$\begin{aligned} DA(t) = \{(a_1,a_2) | (\exists p : p \in Paths(a_1,t) \wedge a_2 \notin p) \wedge \\ (\exists p : p \in Paths(a_2,t) \wedge a_1 \notin p)\} \end{aligned} \tag{A.2}$$

Algorithm 1: DISJANC$(T,(A_1,A_2))$

Require: $IC(a_1) \leq IC(a_2)$

1: $nPaths = |Paths(a_1,a_2)|$
2: $nPaths_1 = |Paths(a_1,t)|$
3: $nPaths_2 = |Paths(a_2,t)|$
4: **return** $nPaths_1 \geq nPaths * nPaths_2$

$$\begin{aligned} DCA(t_1,t_2) = \{a_1 \mid a_1 \in CA(t_1,t_2) \wedge \\ \forall a_2 : (a_2 \in CA(t_1,t_2) \wedge IC(a_1) \leq IC(a_2) \wedge a_1 \neq a_2) \\ \implies (a_1,a_2) \in DA(t_1) \cup DA(t_2)\} \end{aligned} \tag{A.3}$$

$$SI_{GraSM}(t_1,t_2) = \frac{\sum_{t \in DCA(t_1,t_2)} IC(t)}{|DCA(t_1,t_2)|} \tag{A.4}$$

A.4. Adjusted GraSM

Adjusted GraSM makes only a slight change to the original algorithm. Changing the *greater than or equal to* (≥) in Algorithm 1 to a *strictly greater than* (>) in the path number comparison causes the algorithm's behavior to change, and the size of the DCA set is reduced. Supplemental Information Trace 1 and 2 provided a detailed example of the differences between these two algorithms on a small calculation.

A.5. Semantic frontier

The semantic frontier (SF) algorithm was developed to address the potential under estimation of shared information and costly runtime of GraSM. The SF algorithm calculates shared information by averaging the IC of concepts that form the border between the common ancestor set and ancestors unique to each term. We term these concepts the *semantic frontier* because they represent different borders to a shared territory. The SF calculation is outlined in Algorithms 2 and 3. The *target* function in Algorithm 3 returns the destination node of a directed edge. The SF algorithm is based on the concept of a breadth first search (BFS) visitor. A BFS visitor performs actions when certain events in a BFS occur. The SF algorithm constructs a BFS visitor that starts at a specific term in the GO graph and visits all ancestor terms on its way to the ontology root. Anytime a new edge is visited, it is added to the *visitedEdges* set for the starting term. Two sets are constructed in this way (*visitedEdgesT*1 and *visitedEdgesT*2) for the terms of interest t_1 and t_2. Any term *t*that satisfies Eq. (A.5) is a member of the semantic frontier of t_1 and t_2.

Algorithm 2: SEMANTICFRONTIER(T_1,T_2)

```
Anc = ConnomAncestors(t1, t2)
visitedEdgesT1 = ∅
visitedEdgesT2 = ∅
IncidentEdges = An empty Map of vertex v to incident Edges e
VisitedEdgeBFS(t1, visitedEdgesT1, IncidentEdges)
VisitedEdgeBFS(t2, visitedEdgesT2, IncidentEdges)
frontierTerms = ∅
for a ∈ Anc do
  for e ∈ IncidentEdges[a] do
    if (e ∉ visitedEdgesT1) ∨ (e ∉ visitedEdgesT2) then
      frontierTerms = frontierTerms ∪ a
    end if
  end for
end for
return frontierTerms
```

Algorithm 3: $VISITEDEDGESBFS(T, VISITEDEDGES, INCIDENTEDGES)$

Construct a breadth first search visitor starting at node t
On $ExamineEdge(e)$:
    $visitedEdges = visitedEdges \cup e$
    $v = target(e)$
    $IncidentEdges[v] = IncidentEdges[v] \cup e$

$$t \in CA(t_1, t_2) \wedge \exists e \mid target(e) = t \wedge ((e \in visitedEdgesT1 \wedge e \notin visitedEdgesT2) \vee (e \in visitedEdgesT2 \wedge e \notin visitedEdgesT1)) \quad \text{(A.5)}$$

The SF algorithm is an efficient implementation of an algorithm developed by Zhang and Lai [21] that they call exclusively inherited shared information (EISI). The shared information of these two equivalent methods is defined in Eq. (A.6) where $SF(t_1,t_2)$ represents the set of terms in the semantic frontier (or exclusively inherited common ancestors). Although equivalent in runtime complexity to EISI, the SF algorithm uses an edge visitor to construct only the necessary set of terms and descendants. This leads to a reduction in the search space as well as more opportunities for early exit. A side-by-side runtime comparison of the two algorithms was conducted, but neither algorithm showed a consistent and repeatable speed advantage (data not shown). As the size of the GO graph continues to grow, the SF algorithm may offer more of an advantage through reduction of the search space.

$$SI_{SF}(t_1, t_2) = SI_{EISI}(t_1, t_2) = \frac{\sum_{t \in SF(t_1, t_2)} IC(t)}{|SF(t_1, t_2)|} \quad \text{(A.6)}$$

REFERENCES

1. Ashburner M, Ball CA, Blake JA, Botstein D, Butler H, Cherry JM. et al. Gene ontology: tool for the unification of biology. The Gene Ontology Consortium.. Nat Genet 2000;25(1):25–9. http://dx.doi.org/10.1038/75556.
2. Consortium GO. et al. Creating the gene ontology resource: design and implementation. Genome Res 2001;11(8):1425–33.
3. Guo X, Liu R, Shriver CD, Hu H, Liebman MN. Assessing semantic similarity measures for the characterization of human regulatory

pathways. Bioinformatics 2006;22(8):967–73.

4. Jain S, Bader GD. An improved method for scoring protein–protein interactions using semantic similarity within the gene ontology. BMC Bioinf 2010;11(1):1.
5. Wang JZ, Du Z, Payattakool R, Philip SY, Chen C-F. A new method to measure the semantic similarity of Go terms. Bioinformatics 2007;23(10):1274–81.
6. Xu T, Du L, Zhou Y. Evaluation of Go-based functional similarity measures using S. cerevisiae protein interaction and expression profile data. BMC Bioinf 2008;9(1):1.
7. Mulder NJ, Akinola RO, Mazandu GK, Rapanoel H. Using biological networks to improve our understanding of infectious diseases. Comput Struct Biotechnol J 2014; 11(18):1–10.
8. Vafaee F, Rosu D, Broackes-Carter F, Jurisica I. Novel semantic similarity measure improves an integrative approach to predicting gene functional associations. BMC Syst Biol 2013;7(1):22.
9. Pandey G, Myers CL, Kumar V. Incorporating functional inter-relationships into protein function prediction algorithms. BMC Bioinf 2009; 10(1):142.
10. Jiang Y, Clark WT, Friedberg I, Radivojac P. The impact of incomplete knowledge on the evaluation of protein function prediction: a structured-output learning perspective. Bioinformatics 2014;30(17):i609–i616.
11. Lord PW, Stevens RD, Brass A, Goble CA. Investigating semantic similarity measures across the gene ontology: the relationship between sequence and annotation. Bioinformatics 2003;19(10):1275–83.
12. Lord PW, Stevens RD, Brass A, Goble CA. Semantic similarity measures as tools for exploring the gene ontology.. Pacific symposium on biocomputing. vol. 8. 2003. p. 601–12.
13. Pesquita C, Faria D, Falcao AO, Lord P, Couto FM. Semantic similarity in biomedical ontologies. PLoS comput biol 2009;5(7). e1000443.
14. Guzzi PH, Mina M, Guerra C, Cannataro M. Semantic similarity analysis of protein data: assessment with biological features and issues. Brief Bioinform 2012;13(5):569–85.

15. Mazandu GK, Chimusa ER, Mulder NJ. Gene ontology semantic similarity tools: survey on features and challenges for biological knowledge discovery. Brief Bioinform 2016; bbw067.
16. Resnik P. et al. Semantic similarity in a taxonomy: an information-based measure and its application to problems of ambiguity in natural language. J Artif Intell Res(JAIR) 1999;11:95–130.
17. Lin D. An information-theoretic definition of similarity.. ICML. vol. 98. Citeseer.1998. p. 296–304.
18. Jiang JJ, Conrath DW. Semantic similarity based on corpus statistics and lexical taxonomy. arXiv preprint cmp-lg/9709008.
19. Pesquita C, Faria D, Bastos H, Ferreira AE, Falcão AO, Couto FM. Metrics for go based protein semantic similarity: a systematic evaluation. BMC bioinf 2008;9(5):1.
20. Couto FM, Silva MJ, Coutinho PM. Measuring semantic similarity between gene ontology terms. Data Knowl Eng 2007;61(1):137–52.
21. Zhang S-B, Lai J-H. Semantic similarity measurement between gene ontology terms based on exclusively inherited shared information. Gene 2015;558(1):108–17.
22. Tao Y, Sam L, Li J, Friedman C, Lussier YA. Information theory applied to the sparse gene ontology annotation network to predict novel gene function. Bioinformatics 2007;23(13):i529 i538.
23. Sevilla JL, Segura V, Podhorski A, Guruceaga E, Mato JM, Martinez-Cruz LA. et al. Correlation between gene expression and Go semantic similarity. IEEE/ACM Trans. Comput. Biol. Bioinform. 2005;2(4):330–8.
24. Azuaje F, Wang H, Zheng H, Bodenreider O, Chesneau A. Predictive integration of gene ontology-driven similarity and functional interactions. Sixth IEEE International Conference on Data Mining-Workshops (ICDMW'06). IEEE.2006. p. 114–9.
25. Gentleman R. Visualizing and distances using Go. :\ ignorespaceshttp://www. bioconductor.org/docs/vignetteshtml.
26. Mazandu GK, Mulder NJ. Information content-based gene ontology functional similarity measures: which one to use for a given biological data type? PloS one 2014;9(12). e113859.

27. Xu Y, Guo M, Shi W, Liu X, Wang C. A novel insight into gene ontology semantic similarity. Genomics 2013;101(6):368–75.
28. Wang H, Azuaje F, Bodenreider O, Dopazo J. Gene expression correlation and gene ontology-based similarity: an assessment of quantitative relationships. Computational intelligence in bioinformatics and computational biology, 2004. CIBCB'04. Proceedings of the 2004 IEEE symposium on. IEEE.2004. p. 25–31.
29. Yang H, Nepusz T, Paccanaro A. Improving Go semantic similarity measures by exploring the ontology beneath the terms and modelling uncertainty. Bioinformatics 2012;28(10):1383–9.
30. Thomas PD, Wood V, Mungall CJ, Lewis SE, Blake JA, Consortium GO. et al. On the use of gene ontology annotations to assess functional similarity among orthologs and paralogs: a short report. PLoS Comput Biol 2012;8(2). e1002386.
31. Barrell D, Dimmer E, Huntley RP, Binns D, ODonovan C, Apweiler R. The GOA database in 2009—an integrated gene ontology annotation resource. Nucleic Acids Res 2009;37(suppl. 1):D396–D403.
32. Jaccard P. The distribution of the flora in the alpine zone.. New Phytol 1912;11(2):37–50.
33. Breiman L. Bagging predictors. Mach Learn 1996;24(2):123–40.
34. Pesquita C, Pessoa D, Faria D, Francisco Couto. CESSM: Collaborative evaluation of semantic similarity measures. JB2009: Challenges Bioinform 2009;157:190.
35. Altschul SF, Gish W, Miller W, Myers EW, Lipman DJ. Basic local alignment search tool.. J Mol Biol 1990;215(3):403–10. http://dx.doi.org/10.1006/jmbi. 1990.9999.
36. Finn RD, Bateman A, Clements J, Coggill P, Eberhardt RY, Eddy SR, Heger A, Hetherington K, Holm L, Mistry J. et al. Pfam: the protein families database. Nucleic Acids Res 2013;gkt1223.
37. Eddy SR. et al. A new generation of homology search tools based on probabilistic inference. Genome Inform. vol. 23. 2009. p. 205–11.
38. Salton G, Buckley C. Term-weighting approaches in automatic text retrieval. Inf Process Manag 1988;24(5):513–23.

39. Song N, Sedgewick RD, Durand D. Domain architecture comparison for multidomain homology identification. J Comput Biol 2007;14(4):496–516.
40. Croft D, Mundo AF, Haw R, Milacic M, Weiser J, Wu G. et al. The reactome pathway knowledgebase. Nucleic Acids Res 2014;42(D1):D472–D477.
41. Marina Meila.˘ Comparing clusterings—an information based distance. J Multivar Anal 2007;98(5):873–95.
42. Ward JH Jr,. Hierarchical grouping to optimize an objective function. J Am Stat Assoc 1963;58(301):236–44.
43. Brown KR, Jurisica I. Unequal evolutionary conservation of human protein interactions in interologous networks. Genome Biol 2007;8(5):1.
44. Sing T, Sander O, Beerenwinkel N, Lengauer T. ROCR: visualizing classifier performance in R. Bioinformatics 2005;21(20):7881. :\ ignorespaceshttp://rocr. bioinf.mpi-sb.mpg.de.
45. Dietterich TG. Ensemble methods in machine learning. International workshop on multiple classifier systems. Springer.2000. p. 1–15.
46. Collins SR, Kemmeren P, Zhao X-C, Greenblatt JF, Spencer F, Holstege FC. et al. Toward a comprehensive atlas of the physical interactome of Saccharomyces cerevisiae. Mol Cell Proteomics 2007;6(3):439–50.
47. Mazandu GK, Mulder NJ. A topology-based metric for measuring term similarity in the gene ontology. Adv Bioinforma 2012;
48. Mazandu GK, Mulder NJ. Information content-based gene ontology semantic similarity approaches: toward a unified framework theory. Biomed Res Int 2013;
49. Couto FM, Silva MJ. Disjunctive shared information between ontology concepts: application to gene ontology. J Biomed Semantics 2011;2(1):1.

REFERENCES

1. Araujo, I. S; Veit E.; Moreira, M.. A study on the performance of students of Physics users Modellus computational tool in the interpretation of graphics in Cinemática.Disponível in: www.if.ufrgs.br/cref/ntef/producao/IV_ENPEC.pdf. Accessed on: 30/06/06.
2. Bassanezi, RC Teaching and learning with mathematical modeling: a new strategy. Sao Paulo: Context, 2002.
3. Bassanezi, R. Modeling as a teaching-learning strategy. For the learning of mathematics, Vancouver, v. 14, no. 2, p. 31-35, June 1994.
4. BARBOSA, JC Mathematical Modeling in the classroom. Perspective, Erechim (RS), vol.27, n.98, p.65-74, June, 2003
5. BARBOSA, JC Mathematical Modeling:. Conceptions and experiences of future teachers, 2001. Doctoral Thesis. Institute of Geosciences and Exact Sciences, UNESP Sciences.
6. BLUM, W. Applications and Modeling in mathematics teaching and mathematics education: some important aspects of practice and of research. In: SLOYER, C .;

7. Blum, W .; HUNTLEY,I. Advances and perspectives in the teaching of Mathematical Modeling and Applications. Newark: Water StreetMathematics, 1995. p. 1-20.
8. BORBA, MC; MENEGHETTI, RCG; Hermini, HA modeling, graphic calculator and interdisciplinarity in the classroom of a course of biological sciences. Mathematics Education magazine SBEM-SP [São José do Rio Preto], n. 3, p. 63-70, 1997.
9. BURAK, D. Mathematical modeling: actions and interactions in the teaching - aprendizagem.Campinas. Doctoral thesis. UNICAMP-FE 1992.
10. Carmen, A. B; CARVALHO, AMP Starting students in mathematical physics through investigative experimental classes. X EPEF, UEL, 2006
11. FERRUZZI, EC et al. Mathematical modeling as a teaching strategy and learning in higher education technology. World Congress on Engineering and Technology Education. March 14-17, 2004 São Paulo, Brazil.
12. Fiorentini D; Lorenzato, S. Research in mathematics education: theoretical and methodological paths. Campinas, SP: Authors Associates, 2006.
13. Knuth, DE [1969] The Art of Computer Programming, Vol. 2, Seminumerical Algorithms. Addison-Wesley. Reading, Massachusetts, USA.
14. LOZADA, C.; ARAUJO, M, Morrone W; Araujo, MST; AMARAL, LH The importance of mathematical models for physics education in high school. Proceedings of the Third International Symposium on Research in Mathematics Education (SIPEM), Aguas de Lindoia, 2006.
15. NOVAK, JD Learning, creating and using knowledge: Concept maps as facilitation tools in schools and businesses. Lisbon: Maple technical issues, 1998.
16. PATY, Michel Qualitative physics and physical mathematized in the eighteenth century. II International Seminar of Philosophy and History of Science, USP, 2003

17. RAMALHO JR, F; FERRARO, N. L; SOARES, P. A. T. The fundamentals of physics. Vol 3. São Paulo:. Moderna, 2005.

18. Richard, E. C. et al. The didactic relationship, barriers to learning and the theory of Conceptual Fields. In: International Meeting Language, Culture and Cognition, 2003, Belo Horizonte.

19. Rubinstein, RY [1981] Simulation and the Monte Carlo Method. Haifa, Israel.

20. Silveira, F. L; PEDUZZI, L.. Q. Three episodes of scientific discovery: the empiricist caricature to another story "Cad Bras... Ens. Fis, v.23, n.1:. P.27-55, abr.2006.

21. Sousa, cmsg; FÁVERO, MH Conceptions of Physics teachers on problem solving and the teaching of physics. Journal of Research in Science Education, 2003.

22. TEODORO, VD Computational Modeling in Science and Mathematics. Magazine Education and Mathematics: APM, n. 45, 1997.

23. Veit E.; TEODORO, VD Modeling in the teaching / learning of Physics and the new National Curriculum Guidelines for Secondary Education. Rev. Bras. Ens. Fis.vol.24, n.2, junho2002.

24. Veit E.; MORS, PM; TEODORO, VD Illustrating Newton's Second Law in the XXI century. Rev. Bras. Ens. Fis.vol.24, n.2, junho2002.

25. ZYLBERSZAJN, A. Problems in Physics Teaching. Available in: http://server.fsc.ufsc.br/~arden/problkuhn.doc .Access on: 25/06/06

CITATIONS

CHAPTER 4

Konstantina Kourou , Themis P. Exarchos , Konstantinos P. Exarchos, Michalis V. Karamouzis, Dimitrios I. Fotiadis, Machine learning applications in cancer prognosis and prediction, Computational and Structural Biotechnology Journal Volume 13, 2015, Pages 8–17

CHAPTER 5

Ioannis Kavakiotis, Olga Tsave, Athanasios Salifoglou, Nicos Maglaveras, Ioannis Vlahavas, Ioanna Chouvarda, Machine Learning and Data Mining Methods in Diabetes Research, In Computational and Structural Biotechnology Journal, Volume 15, 2017, Pages 104-116, ISSN 2001-0370, https://doi.org/10.1016/j.csbj.2016.12.005.

CHAPTER 6

Tuhin Kumar Guha, Alvan Wai, Georg Hausner, Programmable Genome Editing Tools and their Regulation for Efficient Genome Engineering, In Computational and Structural Biotechnology Journal, Volume 15, 2017, Pages 146-160, ISSN 2001-0370, https://doi.org/10.1016/j.csbj.2016.12.006.

CHAPTER 7

Mohammad R. Mohebian, Hamid R. Marateb, Marjan Mansourian, Miguel Angel Mañanas, Fariborz Mokarian, A Hybrid Computer-aided-diagnosis System for Prediction of Breast Cancer Recurrence (HPBCR) Using Optimized Ensemble Learning, In Computational and Structural Biotechnology Journal, Volume 15, 2017, Pages 75-85, ISSN 2001-0370, https://doi.org/10.1016/j.csbj.2016.11.004.

CHAPTER 8

Paul W. Bible, Hong-Wei Sun, Maria I. Morasso, Rasiah Loganantharaj, Lai Wei, The effects of shared information on semantic calculations in the gene ontology, In Computational and Structural Biotechnology Journal, Volume 15, 2017, Pages 195-211, ISSN 2001-0370, https://doi.org/10.1016/j.csbj.2017.01.009.

INDEX

Index

A

accurately 247
antenna directivity (dB) 12
any of its descendants 286
Area under the curve (AUC) 115
Artificial Neural Networks (ANN) 147
Artificial Neural Networks (ANNs) 110
artificial zinc-finger protein (AZP) 199

B

Bagged Decision Tree (BGD) 249
Bayesian Networks (BNs) 110
best-match average 284, 292
biological_process 286
biological processes (BP) 283

C

Cancer data 110, 121
cellular components (CC) 283
clinic pathologic characteristics 246
Clostridium difficile 217
common ancestors (CA) 286
Compilers 85, 86
Computer-aided diagnosis (CAD) 244
Computer Networks 82, 83, 92
Computer programs 77

D

data communications 81, 83
Decision Trees (DTs) 110
diabetes mellitus (DM) 145
Diabetes mellitus (DM) 144
discovery in databases (KDD) 145, 146
disjunctive common ancestors (DCA) 290, 323

Distributed systems 84, 101
DNA double-strand breaks (DSBs) 191

E

Electronic Health Records (EHRs) 143
Enhanced Green Fluorescent Protein (EGFP) 201
evaluation device distribution 7

F

Feature selection (FS) 250
flexibility 25

G

gene ontology (GO) 281, 282
GO Graph Tool Kit (GGTK) 282, 286

H

hammerhead ribozyme (HHR) 215
Heterogeneous disease 109
High Throughput Technologies (HTTs) 121

I

inferred from electronic annotation (IEA) 283
Instance Based Learning (IBL) 147

K

Kolmogorov–Smirnov (KS) 251

L

Lactococcus lactis 208
Learning Classifying Systems (LCSs) 126

M

machine learning (ML) 145
Machine learning (ML) 110
Magnetic Resonance Imagery (MRI) 144
Mathematical modeling 70, 71, 74
Mathematics 79, 105, 332, 333
mathematization 61
most informative common ancestor (MICA) 284, 286, 290
Multidimensional heterogeneous data 135

N

non-homologous end joining repair system (NHEJ) 190
nonsense-mediated decay (NMD) 196, 204
Nottingham prognostic index (NPI) 246

O

odds ratio (OR) 262
Operating systems 82
Oral squamous cell carcinoma (OSCC) 127

P

Particle Swarm Optimization (PSO) 244, 249, 252
peptide nucleic acids (PNAs) 192
private mobile media (PMR) 3
protospacer adjacent motif (PAM) 200

R

repeat variable di-residues (RVDs) 199
ribonucleoprotein particle (RNP) 203
RNA interference (RNAi) 190

S

shared information (SI) 282
Software Engineering 86, 89, 92
Staphylococcus aureus 202
Streptococcus pyogenes 196, 202
structure-guided endonucleases (SGNs) 192
supported vector machines (SVM) 246
Support vector machines (SVM) 144
Support Vector Machines (SVM) 147
Support Vector Machines (SVMs) 110

T

Tet-response element (TRE) 201
The Cancer Genome Atlas Research Network (TCGA) 122
Treponema denticola (TD) 196

U

ultraviolet (UV) 211

W

Wrapper algorithm 127